Wolves and Wilderness

Illustrations by
Mary Theberge

TORONTO / DENT CANADA

Wolves and Wilderness

John B. Theberge

ISBN 0-460-95829-1

Some of the passages in *Wolves and Wilderness* have previously appeared in 'Alaska', 'Audubon', 'Natural History', 'Ontario Naturalist', and 'Outdoor Canada'.

Printed in Canada

1 2 3 4 5 6 7 8 9 10 Bry 84 83 82 81 80 79 78 77 76 75

TO

Mary, Lawrence, Doug, and the others who, in pursuit of a greater understanding of wolves and wilderness, have shared my campfires, and to the handful of other wolf research biologists whose insight into wolves and their wilderness home has given new hope for both.

Contents

Prologue

For man, life in the northern forest has been a never-ending search. Generation has followed generation; the quarry changes but the search goes on.

Long before Etienne Brûlé found his way to the country of the Nipissings, man was searching the wilderness for the means of life itself. He stalked deer, drinking by the water's edge, snared rabbits on the game trails, speared the trout in shady pools. He sought birchbark for his canoe and limbs of white ash for his bow. To succeed in the search was to live — to fail was to die.

Then came the missionaries, facing starvation and torture as they searched for the souls of men. Father Le Caron celebrated the first Mass in the country of the Hurons, and the light of European civilization crisscrossed the wilderness.

Another hundred years and the freight canoes of the traders were being carried over the portages on their way west. No longer was it necessary for tribal hunters to find white ash to shape into bows. They shifted their search to beaver and received muskets in return; they exchanged four skins for a knife, six for a hatchet, fifty for a gun. Beaver was coin of the realm.

Then came trappers and land surveyors. Not content to let the Indians bring the wealth in furs out of the forests, these men searched for new paths through the wilderness. Famous men, like David Thompson, who was the first white man to descend the

Columbia, and forgotten men, like Lieutenant Briscoe of the Royal Engineers, tried to find a water route between the Ottawa River and Georgian Bay.

Again the quarry changed; the search shifted to the finest stands of virgin pines, both red and white, which would bring more wealth for the markets of the old world and the new. This was the age of the timber barons — men like Booth and McLachlin — the day of the river driver and the shantyman.

Gradually, almost without knowing it, man was becoming aware of a different kind of wealth. He discovered in these forests rewards more precious than beaver or pine. In 1893 an expanse of rocks, trees, rivers, and lakes was set aside as a park — nearly three thousand square miles for the enjoyment of generations to come. It was called Algonquin Park. In it Tom Thomson searched for beauty and captured it in canvases that started a whole new school of Canadian painting.

Today, in this age of changing values, the search goes on. Men and women are finding spiritual and aesthetic satisfactions in this wilderness park. A Boy Scout from a large city tastes his first adventure in the outdoors; a man and wife, surfeited with urban living, find new perspectives as they explore lonely lakes and silent portages.

My story is of a modern search: a search for wolves in Algonquin and the wilderness lands farther north. I have hunted wolves not with gun and trap for traditional reasons, to kill them and nail their pelts to a wall or collect a bounty, but with notebook and binoculars in the interests of modern scientific knowledge. In the early years (told in the first part of the book) I was a student member of the Ontario Government's research team directed by Dr Douglas H. Pimlott. Our objective was to understand the ecology of the wolf — to find out how many wolves there were, whether they achieve a limitation on their population density, and what their impact is on the species they eat. Our search for wolves was conducted mostly by howling for them at night. Later (in the second part), when the importance of daytime observations became apparent, I pursued graduate studies to learn their language — why they howl and what they say. Since then I have watched wolves at their dens and rendezvous sites on the northern tundra and back in Algonquin Park, pursuing the truths about wolf predation and the secrets of their social behaviour.

In my search, I have been joined at various times by other people: Lawrence Moore, a woodsman all his life; Doug Pimlott, whose scientific curiosity about wolves stimulated mine; Mary and Jenny, my wife and daughter, who have shared many exciting encounters with wolves. Together we set out to find wolves. But in our searches we found other things. We found that water was more than just stored-up kilowatts of energy — every stroke of our paddles told us that. We learned that trees were more than saw logs and newsprint — wind, sighing through the firs on rocky points, told us that. We discovered that solitude is a resource worth searching for — every quiet beaver pond told us that. We learned that wilderness has a place in today's living — every time we heard the howl of a wolf or the call of a loon, echoing and re-echoing in the darkness, we knew that we had found the richest treasure of all. For all this, and for the practical purpose of maintaining life on earth as well, we understood the vital message Thoreau left us when he wrote: 'In wildness is the preservation of the world.'

Searching the Wilderness

1
Genesis

Snow swirled in the night sky. Fingers of ice, which had formed since dark, stretched out into the lake.

The forest had struggled free a few days before, but now was caught again in winter's dying grip. Spruce trees, reclothed in costumes of wet snow, seemed subdued by the forces which had held them captive for so many months. After a week of warm weather, winter was making its last thrust.

Our canoe slipped from the shore, its bow shattering the thin film of ice. Paddles sent fragments tinkling over the frozen surface. A few quick strokes, and civilization was left behind. We had set off to meet the wilderness on its own terms.

Somewhere in the blackness, concealed by an alder thicket or at the edge of a lonely beaver meadow, were the wolves. It was the denning season, and at this very moment a new generation was feeling for the first time the fetid warmth of the den and experiencing sensations of pleasure, pain, fear, and hunger. These new lives were the products of centuries of evolution, years of pruning, selecting, and culling. And now, perfectly adapted by time to their role in life, nature's master predators were emerging as tiny balls of fur in the hollow of sheltering rocks.

'What's our bearing?'

Instantly I switched on the light strapped to my forehead. As bowman, I was the navigator. A cone of pirouetting snow met us head on, the granules hissing as they pelted the surface of the lake. Outside the cone's white orbit, the sky seemed blacker than ever.

'Two hundred and seventy-five degrees,' I replied, 'due west.'

'We're too far north,' said Lawrence. The compass needle began to move as he swung the canoe.

'Two hundred and sixty-five,' I shouted.

'Hold it there.'

It was comforting to hear Lawrence's voice and the rhythm of his paddle. I remembered the postscript to the letter I had received in January: 'Your partner has spent twenty-two years in Northern Ontario lumber camps. He could find a portage that would baffle an otter.'

The longer you look forward to an event, the more firmly fixed in your mind is the picture of what that event will be like. Then, when the day finally arrives, how different is the reality! Since receiving this appointment I had formed my picture of wilderness canoe travel by night: calm water glistening in the moonlight; black shadows across the portage; the next lake, then the next; our tent pitched in the first light of dawn. I had resolved to anticipate everything — nothing would come as a surprise; I would never be caught off guard. But now that it was happening, what were my thoughts? The swirling cone of light, the black sky and even blacker water, the snow beating against my face. We had set out alone in the darkness — not along the shore, but straight into the unknown. The future had become the present, and the present was alive. The portage at the end of the lake? The next lake? Dawn? For me, at this moment, they did not exist. Then I heard Lawrence's voice coming out of the darkness.

'Time for a howl. We can blast the sound a fair distance from out here.'

He stopped paddling and, unstrapping the pack at his knees, pulled out our portable tape player. The silence evaporated as mechanical wolf howls reverberated from the surrounding hills. When the echoes had died away, we sat listening, trying to catch the faintest reply. Nothing; only the hiss of the snow. I switched on the flashlight and made an entry in the logbook. '12.43 — Cache Lake. Played taped howl — 10 minutes. No response.'

'We need to know the number of wolves in the park,' the letter had read. 'Wolves howl when they hear other wolves. The "other wolves" will be you.'

I was glad we had the tape player with us. My howl, in spite of much practising, still sounded very unconvincing, and I was certain that any wolf which might be taken in by it would be tone deaf.

'Wolves howl most readily at night,' the letter continued, 'so your searching must be done then. To be sure that packs heard one night are not confused with those heard the next, you must travel as quickly as possible.'

Lawrence started paddling again and I joined in. The snow was beginning to clear.

'Shine your light ahead there, we should be near the shore.'

The beam picked out a line of trees.

'We'll paddle to the right a ways,' Lawrence said. 'I think we're south of the portage. Wait. There's an opening.'

We eased in, along the half-submerged trunk of a pine. The track of a small animal was visible in the soft snow lying on the exposed side of the tree.

'Mink,' said Lawrence. 'Was a time that fella's hide was worth two days' hard work on the crosscut saw. That's before ranching was common.'

Under protest from muscles softened by city living, I pulled myself out of the canoe and onto a log. Back a little way from the shore I noticed a half-torn sign nailed to a spruce tree: BE CAREFUL OF FI . . . Beyond this, a path led into the darkness. Lawrence, with the ease that comes with 'twenty-two years in Northern Ontario lumber camps', slipped the canoe over his head. 'Portage from Cache Lake to Head Lake, eighty-one chains,' the map read. Following the trail was easy enough, since canoe trippers of previous summers had worn it into a broad path. The flashlight's beam picked up tracks made by a more recent traveller: a moose, searching for willow twigs, had discovered the trail where it found the footing good, and each hoofprint had turned up a clod of snowy earth. About halfway down the path its tracks headed off into a stand of cedar, and soon we heard it crashing through the brush.

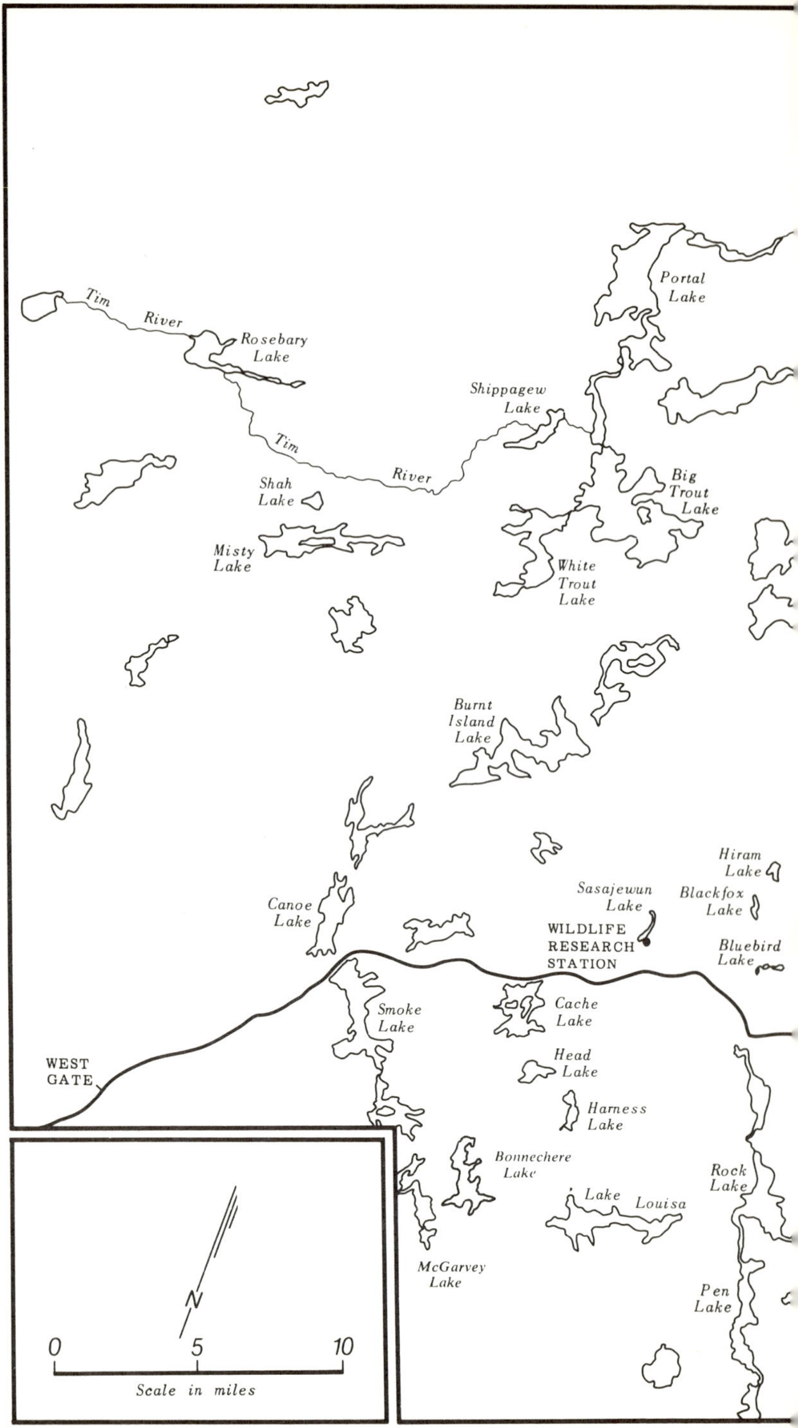
Tim River
Rosebary Lake
Tim River
Portal Lake
Shippagew Lake
Shah Lake
Big Trout Lake
Misty Lake
White Trout Lake
Burnt Island Lake
Hiram Lake
Sasajewun Lake
Blackfox Lake
Canoe Lake
WILDLIFE RESEARCH STATION
Bluebird Lake
Smoke Lake
Cache Lake
WEST GATE
Head Lake
Harness Lake
Bonnechere Lake
Rock Lake
Lake Louisa
McGarvey Lake
Pen Lake
N
0
5
10
Scale in miles

Brent Stn.
CEDAR
LAKE
Petawawa
River
Catfish Lake
Radiant Lake
Petawawa
Philip Lake
Hogan
Lake
River
Big Crow Lake
Lake Lavieille
Little Crow Lake
River
Crow
Proulx Lake
Hardy Bay
White Partridge Lake
North Arm
Dickson Lake
Graham Islands
OPEONGO
East Arm
Annie Bay
LAKE
South Arm
Sproule Bay
FISHERIES LABORATORY
Booth Lake
Sunday Cr.
Fork Lake
EAST GATE
ADMIN. BLDG.
Algonquin Park
Ontario
Wolf Study Area

Moose thrive in Algonquin today chiefly because of man's interference with the environment. Early in the last century, before axes had levelled the stands of giant pines, the sunlight so necessary to life on the forest floor was blocked out by the mammoth trees. Fir seedlings, on which these big mammals thrive, could not grow in the deep shade, so moose could not survive. Occasionally, lightning touched off fires that would burn through the old forest, making way for new growth — first grasses and seedlings, then shrubs and young trees, which provided oases of food for the remaining moose. But despite these periodic fires, in many areas untouched by fire big pines still sealed off the light. When lumbermen came on the scene, the smoke from the logging shanties brought new hope for the dwindling moose population. Soon timber drives down the Petawawa and Madawaska Rivers confirmed the hope, and moose moved in. As a result of more food, birthrates increased as more twins were born and more yearlings produced calves. With the moose came white-tailed deer, another species which thrives best in the new growth after logging or fire. After the large timber had been cut, more of the new growth was maple and yellow birch, foods even more favourable to deer than to moose. The deer population doubled and redoubled, and soon outstripped that of its larger relative.

Life is not easy for deer or moose. They live in constant threat of attack by wolves. Moose have an additional, more subtle enemy: a tiny lungworm, frequently eaten along with aquatic vegetation. This little creature, picked up in the larval stage from snails, burrows its way into the spinal cord and up to the brain, destroying the nerve tissue and finally the moose itself. Hunters call it 'moose staggers disease'. Later I saw a moose afflicted by the parasite blundering into deadfall as though it were blind.

'Want to take it a ways?' Lawrence had stopped, resting the canoe on a stump. He slipped from under the load and helped me settle the centre thwart on my shoulders. Immediately I discovered that the bow of the canoe cut off the beam of the flashlight, which was still strapped around my head. I tilted the bow back and in doing so felt a sudden twinge of pain in a seldom-used muscle; to make matters worse, the weight on my left shoulder pinched my skin. I lowered the bow again and lumbered on, trying to think of something else — like how much lighter a birchbark canoe would have been.

Birchbark. To the Indians who once lived here it was a most versatile resource. They used it for everything from carrying their babies to calling moose. By sewing it with cedar roots and caulking the seams with spruce gum, they fashioned their buckets and kettles. They used it to line food caches and construct their shelters. But most important was its value in making canoes. Before the flames of animosity flared up among the tribes, the Algonquins used to trade their canoe bark with the Iroquois to the south, who . . . Bump! The canoe jarred and slipped from my shoulders. I stepped sideways and my foot caught on a root. Lawrence, hearing the commotion, came back and helped me free myself. 'My turn anyway,' he suggested.

By the time we reached Head Lake the snow had stopped and the moon was shining through the cloud cover. We loaded the canoe, took a compass bearing on the next portage, and pushed off from shore. Halfway across the lake we paused to play another wolf howl; still no reply.

Although the portage between Head and Harness Lakes was the same length as the previous one, it seemed much shorter. I had accomplished the trick of balancing the canoe in just the right spot on my shoulders and began to enjoy the exertion. It was turning cold and a stiff breeze off the lake bit through our heavy clothing, so I was glad when Lawrence suggested that we stop and build a fire. He filled a billy of water and hung it over the flames; then, seeing me pull a loaf of bread from the top of our food pack, he protested.

'Can't eat that till the water boils. And don't watch it, neither. Here, help me get some more dry wood.'

While I was searching for firewood, I wondered about the origin of these fragments of timber-shanty lore with their colourful taboos and predictions. Hewn from observations and experiences in the bush, shaped with a mixture of realism and romanticism, smoothed around countless campfires, they remain today as curios of a bygone age. Wild violets, made into a syrup, will cure rheumatism of the right wrist. Squirrels predict a cold winter by storing extra food in the fall. Eating beaver assures one of the ability to work hard.

As night retreated behind the circle of light from our campfire, I recalled stories of adventure that had fascinated me as a boy. Here, on the shore of this lake, the setting was pure Jack London, except that we were not being watched by the gleaming eyes of half-starved wolves glaring at us from the edge of the circle. In spite of my intentions to be prepared for anything, I scanned the darkness, nervously wondering if perhaps wolves had heard our last howl and right at this moment, were out there, just beyond the flickering half-light — watching.

I had read J. W. Curran's book, *Wolves Don't Bite*, and I was quite convinced of the truth as well as the logic in his remark that 'Any man who says he's been et by a wolf is a liar.' I also knew that authentic records of people being attacked by wolves were extremely rare. But in spite of logic, fact, and science, there were times that first summer, given the right setting, when an uneasy thought would plague me. Who, I reasoned, of all people, stood the greatest chance of proving Curran's statement wrong? Who might become the first authentic wolf victim? It would be Lawrence for sure — I would be up the nearest tree.

By the time we reached the next portage, the moon had disappeared. Large flakes of wet snow were again filling the night sky, and spruce trees along the trail sent down icy showers as our packs brushed their snow-laden limbs. From a high point I looked back to see a strange sight. The shadowy form of the canoe seemed to be floating upside-down along the trail without any means of locomotion. The darkness, a dip in the path, and Lawrence's short legs had combined to create the ghostly illusion. I wondered why, after all his years in the bush, this wise forest pioneer had decided to spend another season of toil under a canoe. To me, it was an adventure, a taste of new experiences. For an old-timer like Lawrence it could only have been a replay of the past. Or was there another reason? Could it have been that in reliving these memories he found enough happiness to make the effort worthwhile?

'What's the trouble? Lose the trail?' By now Lawrence had halted beside me.

'No. Just taking a breather.'

'Lean over when you want to rest. Get the weight of the pack off your shoulders.'

When we were out onto the next lake we tried another howl, but again our only answer was the silence and the lapping of the water along the sides of the canoe. By the time we reached the far shore, the first touch of dawn was etching the horizon and a white-throated sparrow was singing nearby. What a cool, fresh song — the voice of the spring woods. It sang defiantly, as if challenging the late winter snow that had laid claim to the forest.

In the city, the first light of day strikes evenly, like lights going up on a darkened stage. But in the woods, daylight comes more discriminately — touching, one at a time, each hill, tree, and rock with lights and shadows, as an artist applying the finishing touches to a painting. We saw snow-laden conifers touched with gold; then the bare branches of maples and birches became shaded in; finally rocks and hillsides were highlighted with blues, greys, and streaks of pink. A winter wren wove through its long involved song. A grouse drummed, and I almost believed I could sense its annoyance at having its drumming log covered with snow.

We halted by a pond, buried in high hills. Trees crowded to the water's edge, and, except for a few tamarack, there was nothing in the composition of the forest to indicate the pond's existence. It was one of nature's beautiful blemishes, too small to make any significant change in the vegetation that surrounded it. But how overpowering this spot seemed! When man is stripped of his bulldozers, chain saws, and explosives, and his equipment is nothing more destructive than a paddle, he is humbled by the silent forest. Human beings need a measure of egocentricity to exist in our impersonal cities, but in the wilderness conceit has no meaning. Were we really so much more important than the grouse that drummed or the wren that sang? At that moment we were caught up, just as they were, in the spell of a spring morning. They, not we, were the real tenants of the woods. They did not need a long list of things to clothe, feed, and shelter themselves, or to make it possible to travel through the wilderness. They were naturally adapted to their environment; we were artificially adapted to it by intellect rather than by selection.

We unrolled our sleeping bags at the base of a large white pine, with branches that had sheltered the ground from the snow. Before I dropped off to sleep, I watched a chickadee busily exploring the limbs above my head and a nuthatch hitching himself up the trunk, examining each crevice in search of food.

By the time I awoke, the warm afternoon sun was shining across my sleeping bag. The snow had melted and winter's last fling was over. Lawrence was standing on the edge of the lake, casting a lure and retrieving it with short jerks. A trout lay on the rocks at his feet. When he saw me stir, he brought the lure to the tip of his rod and walked up from the shore.

'Good spot there,' he said. 'It drops off deep. Like to take the rod and give it a try? I've had a couple of dandy strikes. I'll fix up the fella I caught for our breakfast.'

The forest was in a relieved and happy mood, as a hostess speeding the departure of a guest who has overstayed his welcome. This time, winter would surely be gone for good. Under the leafy litter, pale shoots of white trillium and columbine would be sprouting. Soon new guests would be made welcome as the migrant birds returned, filling the woods with their songs. Water trickled out of rock crevices, gathering in every low spot. The perfume of spring was blended from the smells of decaying leaves, warm wet earth, and pine needles.

Late that afternoon, after crossing Louisa and McGarvey Lakes, we settled ourselves comfortably against a hemlock tree to wait for dark. From where we sat we could watch the sun dipping below the ragged shoreline and the shadows lengthening over the water. We tried a few wolf calls, vocally this time, imitating as best we could the recorded howls. The sound bounced around the lake, then silence.

What turmoil had once convulsed this spot! Ice blocks toppled from glacial walls as great frozen masses advanced and retreated. Spray showered a hundred feet in the air when splinters of ice the size of ten-storey buildings splashed down into silt-laden waters. Fierce gales from retreating ice fields scoured the granite wastes. No trace of life, plant or animal, was left. All this happened here in Pleistocene times, more than ten thousand years ago — just yesterday on the geological time clock of the universe.

Slowly, ever so slowly, life crept back. In this chaotic wasteland the requirements for existence had not been destroyed: carbon dioxide and water for photosynthesis, minerals in the rocks for nutrients, and oxygen — all fired by the sun's energy. The same sun that was guiding the life processes of the hemlock against which we were resting had once, in Pleistocene times, encouraged pioneer plants, animated rock-growing lichens, and given vitality to mosses clutching at the granite with eager rhizoids. As they died, they formed organically rich soil, a substrate for grasses and ferns. The rootlets of these, in turn, combined with the soil to hold moisture and make possible the growth of shrubs. When conditions were right, seeds of aspen and birch were carried here by the wind from land which had lost its mantle of ice long before. But growth of the first trees did not mean the end of the process. They were replaced first by open, later by dense spruce forests, and even later by hardwoods. Today the forests are a composite of the past: spruces in the cold wet lowlands, maples, pines, and yellow birch on the higher and warmer till soils, jack pine on the well-drained sands left from the melting of the ancient ice.

'Wolves! Quiet!'

With hands resting on the gunwales of the canoe, we held our breath and listened. The moon had disappeared behind a cloud bank which had drifted in from the northwest. The night was almost over, and our canoe was rocking gently on the waters of Bonnechere Lake. Wolves, a mile or more away, were answering our last calls with long, mournful, deep-throated howls overlaid with sharp yapping and barking. This was success. We had found our first wolves and they were answering us. Had we fooled them or did they really know? A chill crept up my spine and into the back of my neck as I listened to the indescribable beauty of that wild chorus and realized that it was their answer to our vocal howling, not to the mirrored sound of the tape player. We had been accepted by the creature which, above all others, symbolizes the spirit of the wilderness. When the howling had stopped, we sat for a long time listening for more. What were the wolves doing? Were they coming closer? Had we unknowingly invaded their territory? One of the only partially solved questions about wolves is the extent of aggression between packs in defence of their territories. Normally, wolves defend their home ground by passive means — by howling or leaving scent from urinating. In rare instances, however, fights have been witnessed. Fighting is much less common among animals than the milder forms of defence known as 'ritualized' behaviour, which warns others of the aggressor's individual rights.

We paddled to the end of the lake and howled again. Immediately the wolf choir responded.

'Straight in from the big spruce on the skyline,' Lawrence indicated. 'Must be at least five or six of them.'

Wolf packs in summer are usually family units: the breeding pair, yearling animals not yet of breeding age, and older adults. Later observations confirmed that a wolf pack in Algonquin rarely consists of more than eight fully grown wolves. What limits the size of the packs is still a mystery; perhaps the answer is found in their rigid social hierarchies, working in some way we have yet to unravel.

As I was logging the compass bearing, the howls ended abruptly. For what seemed like minutes the echoes reverberated in our ears and we sat motionless until everything was silent again. Still under the spell of these wild sounds, we picked up our paddles and headed towards a rocky point to make camp.

Sleep did not come quickly after our wilderness concert. The night seemed especially short, for at dawn the wolves howled again — from the same direction as before. If they had investigated us they must have done so quietly, detecting the dreaded scent of man, the one scent from which their instinct would tell them they must run.

'Might be a storm coming up,' Lawrence commented as we zipped ourselves into our sleeping bags. 'Sudden change of temperature and that bank of clouds. Besides that, seen my first black fly. She's gonna rain.'

Like most woodsmen, Lawrence had his own philosophy about many things — elemental things like the weather, or when the trout would bite. 'Fishin's no good till the leaves on the lilac are as big as a mouse's ear,' he would say.

The storm came the next afternoon. Secure in our well-anchored tent we watched black clouds race down the lake. All morning, lightning had been skulking about the sky with constant diffused flickerings, and the sides of our tent heaved gently in the sultry air. Soon we could see rain driving over the water towards our camp, and we braced ourselves for the onslaught. The storm hit with a violence that billowed the tent and threatened to carry it away. The balsams between us and the lake bent under the force of wind as it tore through their branches. Lawrence was pleased that his prediction about the storm had come true — I could tell by the way he held his cigar uplifted in his mouth.

Although a small man, my companion always seemed uncomfortable in a tent. As soon as the violence of the storm had passed and it was reduced to a mild shower of raindrops filtering through the balsams, he went in search of dry firewood. In a few minutes he was back with an armful of pine knots which he deposited on the floor of the tent; then, pulling his bait box from his pack, he examined each lure affectionately. When he came to the 'ottertail', he held it up, looked at it more closely, relit his cigar, and waited for me to suggest that a bit of fishing might be profitable after the rain.

'Think the fishing would be good about now?' I asked.

'Maybe,' he replied. 'I seen a few jumping.'

We pushed the canoe out, and Lawrence paddled while I trolled. A loon called, but nightfall was too close for any of the forest creatures to revive their interest in any daytime activities. It was as if they considered this to have been a lost day and had decided to wait for dawn. In the quietness we could hear trout breaking the surface across the lake, so we headed towards the splashes.

Limnologists call this period in the annual temperature cycle of a lake the 'spring overturn', the time when the stable temperature gradient set up under the ice breaks down. All winter the water lying just under the ice is near the freezing point, while that at the bottom, where the heaviest water sinks, remains four degrees Celsius warmer. These constant temperatures result in two distinct layers, with no mixing of surface and bottom waters. When the ice goes out and the temperature of the surface water rises the magic four degrees, the water is free to mix. Lakes often turn brown in the spring from bottom sediments stirred up by this mixing. Later, in the summer, a stable temperature gradient is again established with the warm water circulating at the top and the cold water lying below. During these changes, trout seek the water level most favourable to them. Preferring cold water, they cannot remain long at the surface during summer, but in the spring they can move about at any level.

'If you hook one, keep your line tight.' But Lawrence's advice was not needed. We tried deep trolls and shallow trolls; we fished close to shore and well out in the lake. Finally we gave up. Just as we neared our camp, a trout boiled to the surface behind the canoe. I quickly let out my line, only to have it snag on the bottom. The fish boiled again, mocking our bad luck.

In the evening, as a wood thrush sang nearby, we took down the tent and packed our belongings once again. That night we paddled out to the highway at Smoke Lake. The miles of blackness gurgled under our bow, slipped beneath our keel, and were lost in our wake. A truck was waiting for us at the road. We roped the canoe to the top and were back in the world of man once more.

At the Research Station, I plotted on a map the location of our first wolf pack. Later in the summer we would make that trip again, but for the moment countless other unsearched swamps, hardwood ridges, lakes, and portages beckoned to us. This trip, as it turned out, was the beginning of my years of research into the ways of the wolf.

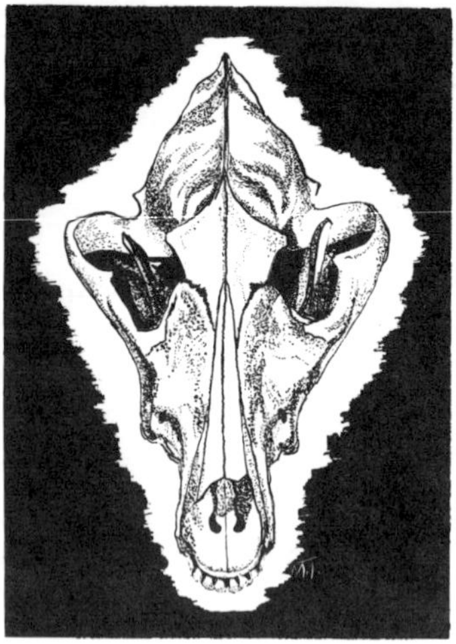

2 Wolf Wake

Smoke billowed skyward. Flames licked at the hardwood logs, then leaped six feet into the sky. A stench, strong enough to topple a bull moose, polluted the woodland air.

The spring ritual had begun, the 'sugaring-off' of the wolf biologist. All winter, the carcasses shipped in to obtain the wolf bounty had accumulated at Research Headquarters. Over a thousand heads had been carefully removed from bodies and stacked in a huge walk-in freezer. In the final diagnosis, they would tell the scientist whether they came from a wolf, a coyote, or a farmer's dog. (It was not unusual to find that the mortal remains of man's best friend had been sold by an unfeeling owner for a few dollars of bounty.)

The wolf heads would be used for research. They would delineate subspecies and provide evidence of hybridization. Since this information could only be read from the white skulls, heads had to be stripped of their trappings of hair, ears, eyes, and brains. Before Lawrence and I could escape to the wilderness road again, we faced this grisly chore.

A few years before, I had pioneered this ritual and served my apprenticeship boiling heads in a big soup kettle over a Coleman stove. But, as the preparation for cleaning made it necessary to boil the heads for twenty-four hours, and this year we faced a pile of over a thousand heads, the process was much too slow.

With our thoughts back in our canoe and breathing the fresh, clean, wilderness air, we easily convinced ourselves that for this year's cookout we needed a bigger boiling pot. We searched the camp and tried out a number of vessels. A discarded wash boiler leaked so badly that it threatened to put out the fire. The cook's garbage pail looked a lot bigger, but it didn't hold any more heads than the soup kettle. We had to abandon these before the first batch of heads was nearly ready.

It was then that Lawrence remembered the bathtub. A bathtub, here in the forest where one wouldn't expect to find an even more basic article of personal plumbing? I couldn't have been more astonished if he had suggested borrowing a lake basin from that mythical giant, Paul Bunyan. Fortunately, I had the good sense to say nothing, and to take off after Lawrence through the woods.

Five minutes of trail-breaking and there it was. A beautiful, king-sized bathtub, lying half-concealed in the underbrush. This was no swab, soap, and splash contraption that only holds one piece of you at a time, but a genuine oversize, railway hotel bathtub, big enough for a six-footer like me to luxuriate in.

Here, at last, was the perfect wolf-head boiler, and we lost no time in hauling it to a clearing, plugging the drain hole, and suspending it over a fire pit. The tub held an unprecedented one hundred heads, each wrapped in cheesecloth so that any teeth which might fall out could be reassigned to their rightful owners.

We built our fire and tended it day and night until the gruesome job was finished. As darkness deepened, we must have appeared grotesque, seated in the flickering firelight beside a great cauldron in which our wolf broth hissed and spluttered.

Hours later, as the chill of approaching dawn caused us to draw closer to the fire, Lawrence recounted the 'Tale of the Tub'. Our container was no ordinary piece of bathroom furniture: once upon a time, its contents had been nobler than what it held now.

The story goes back to the days when grand tourist lodges were permitted to operate in Algonquin Park. Then the tub had been the property of the Grand Trunk Railway, providing a citified luxury in their famous Highland Inn on the shore of Cache Lake. At the time, the railway was the only route to Algonquin, and the park was the playground of the well-to-do.

In the summer of 1919, Canada was decked out to welcome a royal visitor. The Prince of Wales was visiting the North American Dominion, and in the course of his elaborate tour he rode on the Grand Trunk through Algonquin Park. Legend has it that on reaching the Highland Inn after his long dusty train trip, the Royal Visitor felt the need of a refreshing bath. This, according to Lawrence, was the tub's most glorious moment, the high point in its long career. From that day, its fortunes had gone steadily downhill and it was now serving its final use to mankind. The hotel had long disappeared; expired leases on private property were not renewable under a government's wise policy. With the closing of the Highland Inn, the unwanted bathtub became government property and was eventually moved into the woods near the Research Station, where it lay unused until we rescued it for our useful but inglorious purpose. Lawrence finished his story with tears in his eyes — a gust of wind had blown the stench in his direction.

While we felt mildly sorry over the fate of the bathtub, something else in our unreal surroundings had suffered a more melancholy fate and caused us real sadness — the wolf heads. We forced ourselves to be callous and regard them as if they had never been alive. But they had. Each of them had been the perceptual centre of a noble animal. In those brains we so unfeelingly scraped out had been neural pathways we barely understand. All the highly evolved social behaviour of wolves, perfected over aeons of time, had once been co-ordinated there.

How could anyone sentence such a superb animal to death? Perhaps admiration for one's fellow creatures, even for those with which man can easily identify himself, is beyond the conception of some people; they are only able to understand the man-animal relationship when they exert their complete domination by killing. Each head in that pile represented a flare of human cruelty: the shallow triumph of a man with a gun or trap, or even a snowmobile or plane, over a wolf. Each head fired the image of the mighty hunter in some member of the 'superior' species, and placed twenty-five dollars of public bounty money in his wallet. At the same time, each head symbolized ignorance of the predator's role in biotic communities, a lack of ethical restraint that has eliminated the wolf from over ninety per cent of its former range in the United States and from much of Canada's southern fringe.

In some places, specific conflicts with livestock require local wolf control, but only a few heads in that pile were the victims of necessary destruction. The executioners had been brought up in the Little Red Riding Hood tradition to hate wolves. Some may even have flattered themselves into believing they had done a service to mankind and basked in the brief fame of an exclusive photo story in the *Smalltown Bugle*.

The law that bequeathed a thousand wolf heads to Lawrence and me had not been modified to suit the changing times; even before our years of boiling heads it had petrified into a sort of living fossil. It had come into existence in 1793 as the tenth act of the first Parliament of Upper Canada and was designed to 'encourage the destruction of wolves and bears'. When the machinery of lawmaking first came to Upper Canada, the statute makers faced the task of providing the citizens with basic rights which hitherto existed only by good faith. The menace of wolves was of such concern to the early settlers — doubtless a justifiable concern, since farms hacked out of the wilderness left livestock vulnerable to predation — that the Wolf Bounty Act was passed. Ironically, this tenth act was thought important enough to follow legislation which, on the law books at least, put an end to the menace of illegal wives; the ninth act had legalized matrimony entered into without clergy.

For 180 years, with only slight alterations, the wolf bounty remained on the statute books of the Province of Ontario. Then, in 1972, impelled by the pressure of public opinion, a timid provincial government announced its intention to discontinue the wolf bounty. With political astuteness, and concern for the handful of citizens who for pleasure or profit liked to shoot wolves, the reason given for rescinding the cruel and useless act was not ethics but 'reducing public expense'.

'Pass me another head.' With a long stick I fished a steamy mass from the kettle. 'Thanks,' murmured Lawrence, as he began to strip off the fur. The first skulls had been interesting to examine; there was a thrill in discovering hidden fissures, condyles, and foramens. But after the first dozen or so the discoveries became commonplace.

'Big fella,' said Lawrence, holding a sopping head at arm's length. 'Bet he pushed a hundred pounds when he was healthier. Likely came from the far north, up around James Bay somewheres.'

I looked at the drooped and dripping ears, which, a few months before, this wolf had used to communicate with his pack members. When he held them forward, he presented a threat to others: 'Stay away from my share of the moose carcass.' When he laid them back, he communicated submission, probably rolling over on his back at the same time, and presenting his vulnerable belly to the dominant wolf. With gestures like this, actual fighting is avoided. A wolf maintains his social rank by showing well-understood signals to those above and below him in the hierarchy.

'Reminds me of a story I heard an old fella up north tell,' said Lawrence. After a long pause, while he lighted the stub of his cigar, he began.

'Took place somewhere near here — that's how these tales always start, though it usually ain't true. Old trapper had been slaving away most of the winter, livin' by himself in a lonely cabin. Come March, he had a good pile of fur, and one day he hitched up his dog team to take his catch to the Fort. She was snowin' pretty lively when he shoved off, and his sled was loaded: provisions for a week on the trail, two hundred pounds of dog food, fifty beaver pelts, thirty-five otter pelts, ten wolverine, eight foxes, a pile of muskrat fur, and half a moose for the factor. First night on the trail, just as he was settlin' into his sleeping bag, he heard a wolf howl — long and low and kind of hungry, but a long ways off. He thought nothing of it and went to sleep.

'Next day he made good time, covering about thirty miles, and that night he heard more wolves — a whole pack, howling, snarling, and sounding very hungry. He unlashed his rifle and set it where it would be handy; and since he wasn't overstocked with ammunition, he began to feel a bit edgy. Next day he whipped his dogs into full speed, hoping to leave the wolves far behind. But to his horror, that night, they was there again, at least fifteen of 'em. He built his fire up high, and rested his rifle across his lap. A set of eyes. Bang. A snarl. Four wolves ran out and dragged the dead wolf into the trees, eating him up in a flash. He shot two more during the night, and at the first light of dawn lit out in a fair frenzy, whipping the dogs with one hand, half running, half riding on the sled, shooting over his shoulder when a wolf came close enough; it was life or death. The wolves were getting bolder, knowing that he couldn't aim too well on the run. Then he suddenly remembered the half moose; that would slow the wolves down. He quickly unlashed it and left it behind in the snow. The wolves hardly broke stride; they gobbled it up and kept on coming. The dog food went next, and the wolves thoroughly enjoyed that. He began tossing his own grub overboard — beans, strawberry jam, tea, sourdough. They didn't bother to make a fire and cook them things up right — just bolted them and kept on coming. The poor fella that got the jam never lived to digest it; the other wolves thought it was blood and cleaned him up just like that. Down to nine wolves now, but it only takes one set of fangs to do you in.'

Lawrence struck a match on the heel of his boot and relit his cigar.

'By then he was only a few miles from the Fort. The fur went overboard; he was sobbin' while he done it. They ate that, too. The old fella was near exhaustion, and that's when one sled runner hit a stump and over it went in a heap of dogs and harness. Picking himself up out of the snow, he started running. The wolves fell on his dogs and cleaned them up — not a scrap of hair left — all seven dogs, gone. Bang. Eight wolves left. Bang, bang; six wolves left. The Fort was real close now; he could see it on the hill ahead. As each wolf fell, the others ate it and kept on coming. Bang; five now. The trapper was staggering — could hardly keep his balance as he ran. Bang; four. Bang; three. Only a few hundred yards now. Bang; two. It was uphill and hard going. Bang; one. The snow was deep, but he was almost there. Click. No more ammunition. He threw the rifle down. The wolf was almost on him. But then the strap of his snowshoe broke. He half crawled towards the stockade. His breath came in gasps.

He couldn't . . . quite . . . make it. He fell forward in the snow. One second, ten seconds passed; a minute. Nothing happened. Two minutes. He was still alive. He lifted his head — and there, not six feet from him, was the wolf lying dead.'

A long pause, broken by the crack and snap of the fire and the hissing steam rolling up from the boiling mass of wolf heads. Lawrence waited for me to ask why. 'Why?' I ventured finally.

'You'd be dead, too, if you ate provisions for a week, two hundred pounds of dog food, fifty beaver pelts, thirty-five otter pelts, ten wolverine, eight foxes, a pile of muskrat fur, half a moose, seven dogs, and fourteen wolves.'

The end of Lawrence's story coincided with the dinner bell. We piled more logs on the fire and started up the hill to the main camp. There were twenty of us at the Research Station, all working on various projects from sampling blood parasites of local birds to collecting deer droppings. Much of our understanding of the ecology of the transition zone between the coniferous forest in the north and deciduous forest farther south has come directly from such studies. We have learned, for example, that certain species of black flies never fly over the land, but pass their parasites to the blood of animals that frequent the water and invade their environmental niche. We have discovered that mosquitoes are attracted to us primarily by the carbon dioxide we expel; that white-throated sparrows recognize the songs of neighbouring territory holders and do not attack them, but will attack a stranger. The relationship of beavers, otters, and fishers to their environment has been intensively studied in this outdoor laboratory.

During our head-boiling days, Lawrence and I were ostracized by the others at the camp. Nobody would sit near us at the dining table. We were left alone at the far end of the long board, except when our companions shouted at us to pass the butter or send down the Worcestershire sauce. Sometimes the comments became more direct when someone would ask, 'What you been cooking today, Jack? Whatever it is it sure smells terrible.' And Old Jack, the cook, would laugh and say, 'That smell sure ain't comin' from my kitchen!'

While the others pitched a game of horseshoes after dinner, it was back to the wolf heads for Lawrence and me. The afternoon dragged on with talk of the fishing we would do in Hogan Lake and the rapids we would shoot on the Petawawa.

Four o'clock and time to feed the captive wolves, a welcome excuse to get away from our unpleasant job and to see a few living wolf heads for a change. The fifteen wolves at the Station had been kept in captivity since they were pups and were well adjusted to such a life. Although it may seem cruel to confine these wide-ranging animals, captive wolves have proved to be extremely useful in helping man to understand the species better. One of them probably contributed more to the creation of a new image for these much-maligned predators than any other wolf in history. Besides providing us with an excellent insight into wolf behaviour, the antics of this wolf and his companions formed much of the factually entertaining material for R. J. Rutter and D. H. Pimlott's book, *The World of the Wolf.* These wolves appeared on many occasions before both live and television audiences, and inspired the idea that we might find wild wolves through tape-recorded howls which have since reverberated throughout every corner of Algonquin Park. They became spokesmen for the plight of their species, and their howls, heard on radio and television and records in millions of homes, have won over the hearts of listeners in Canada and the United States.

Two cans of dog food, a pile of fox pellets in each trough, a bucket of water, and some affectionate neck scratching — that was the daily allotment of the captives. Every wolf received the same rations — everyone except Yukon. He wanted no part of man's friendly advances. Sometime in his early life, someone had broken faith with him. He had come to the Research Station fully grown, from the southern Yukon where he had been a captive all his life. He feared man and looked upon him as an enemy. When we fed Yukon, we were cautious, being careful to keep facing him. At the other extreme was Big Grey, the prime subject of the wolf communication studies I made later. He accepted me as part of his social unit (in the wild, that would have been his pack), and displayed submissive behaviour — ears and tail down while he rolled over on his back. As affectionate as the gentlest household dog, he begged to be petted.

Evening, and Lawrence and I were back at our head boiling again, laying in a supply of wood before dark. From my Boy Scout days I remembered the cardinal rule against cutting live trees for firewood. They are more difficult to burn, and besides, I had been taught that we desecrate the forest by taking away living material. Since scouting days I have come to realize that a dead tree is alive too, harbouring carpenter ants, grubs of beetles, and fly eggs, and playing host to woodpeckers pounding on the hollow wood. The tree will fall and eventually add its nutrients to the soil or become a nurse log for new seedlings. Can anything in a living ecosystem really be called dead?

Lawrence was skilful with an axe, striking cleanly and making chips fly. Logging had been his life, and he had made good at it. We soon had a pile of spruce logs alongside the bathtub and settled down to cleaning skulls. As we were anxious to resume our search, we worked late that night. About midnight an owl came to supervise; from its perch in the shadows outside the rim of light it serenaded us with its hollow music.

By dawn we had carted our last load of skulls to the shed where they would be boxed for shipment to the laboratory in Toronto. The wolf wake was over. We were off again in search of live wolves and to reassure ourselves that we had not administered the last rites at the funeral of an entire species.

3
The Tim

The river is the forest's Main Street, the street where skilled artisans mould and shape links in the chain of life. Muskrats burrow into its banks, beavers alter its course, mergansers build homes along its margins. In the river's chemical factory, minerals absorbed from bottom sediments produce nourishment for tiny organisms, which in their turn provide food for trout. Trout attract kingfishers, herons, and ospreys. In the same way that life is attracted to a city's main street, it is drawn to this narrow, pulsating, woodland thoroughfare.

Beneath hardwood ridges, buried deep under megatons of earth, compressed against bedrock or trapped in rock-sealed caverns, lies a great store of subterranean water. Gradually, some of this water percolates downwards into the valleys, sidetracking to avoid roots of maple and birch that reach to draw it in. Then, along some gravelly depression, it breaks through the earth's surface. First a rivulet, then a stream, finally a miniature torrent gushes its way into the deeply grooved channel. Now it is the Tim River — the route to Shippagew Lake, the Petawawa, and the Ottawa River. Along this watercourse, for hundreds of miles, life spawns, swims, floats, dives, and wades.

Until the day a logger reported seeing signs of a wolf den near Rosebary Lake, to me the Tim was just a thin blue line on the map. Later, as I looked down on it from the single-engine Beaver, it looked like an aimless right of way hacked out of the forest by some fanatical road builder, obsessed with a compulsion to avoid every tree, rock, and knoll which stood in his way. Soon it would become a highway for our canoe, a road along which we would bump and jostle as the water swelled over hidden rocks and frothed against the impasse of uprooted trees.

The pilot brought the plane down on Rosebary Lake and helped us unlash our canoe from the pontoon strut. With a wave of his hand he taxied upwind and the aircraft disappeared over a line of trees. Lawrence and I put in the time while we waited for dark by paddling slowly around the lake, searching for signs of wolves and making plans for the night's work.

Wolves, like most other forest dwellers, are attracted to water highways. Along these busy waterways they sometimes surprise a deer, wading belly-deep in an effort to escape the savage attacks of black flies, or come upon an unwary beaver, too intent upon his carpentry. Later we discovered that wolf packs frequently use lakes and rivers to mark territorial boundaries. Although wolves are good swimmers they don't like getting wet, and often go miles out of their way to cross a stream at a beaver dam.

It is said that Rosebary Lake was once important to the Indians. Along its shores they gathered rocks which, when pulverized and made into a paste, provided them with their brilliant yellow war paint. From a pebble-strewn beach on the west side we picked up a dozen or more stones which appeared to contain aluminum oxide, but they were poor lacklustre specimens. No doubt the brightest had long ago been collected and used to adorn the faces of Algonquin warriors, fighting to defend their trade routes against invading Iroquois.

To enter the Tim from Rosebary Lake we had to ride the current over a beaver dam. The river was in spring flood, and heavy rains would keep the water high until mid-June. Black clouds in the north threatened a downpour as we set out on the river, a thirty-foot-wide slash twisting and turning in front of the canoe. The way ahead led us through thick coniferous bush, which gradually gave way to wide valleys with marsh grass shoulder high on each side, and hills spreading away in the darkness.

The marshy area presented a strange and startling sight. In

front of us, as far as we could see, were thousands of miniature lights, flashing on and off in the darkness. These fairy lamps glowed brilliantly over the grasses, then stopped abruptly at the water on either side, leaving the main channel a ribbon of blackness. The course of the river was illuminated with fireflies, their greenish-white phosphorescent flashes pricking the night. By standing up in the canoe, I could trace the meandering river through the marsh to where it disappeared into the forest beyond.

Had there been wolves here recently? We stopped paddling and sent a howl resounding over the insect-lighted marsh. Months earlier, when the river was frozen, wild cries like these had been familiar sounds in this valley. Government aircraft patrols had spotted wolf tracks in the snow, the animals preferring the easier footing of the marsh to the icy crust in the deep woods. In winter, wolves may have paused here to scratch at a frozen muskrat burrow, but it would only have been a halfhearted effort, the kind of unconscious gesture humans make when they look in the coin box of a pay phone without any real expectation of collecting any money. The wolves knew that their food wasn't to be caught that easily. It was securely bedded down in conifer thickets, well away from windswept marshes.

In winter, wolf packs may range across a hundred square miles and may travel twenty or thirty miles a day. But when the ice goes out of the rivers and lakes, they are forced to find new travel routes: hardwood ridges, portage trails, and logging roads. Then their range is reduced to twenty-five miles, or even less. This marsh may have heard no wolf howls since the previous winter, and may well have lain fallow without them throughout the summer until they came again the following winter. No matter, there would be other music: the booming choruses of bullfrogs, the rhythmic lashes of whip-poor-wills, the euphony of mergansers, the squawking of herons.

All went well as we rode the water highway over logs and rocks that later in the year would require much manoeuvring to get through. All went well until just before midnight, when we felt the drizzle on our faces. We paddled faster, hoping to reach the first portage before the rain set in. That was our mistake.

We rounded a bend and were suddenly caught up in the rushing swirl of rapids. The canoe lurched forward, then wedged itself between jagged rocks. As if timed to add to this moment of misfortune, the full force of the storm broke. Thunder crashed and lightning silhouetted every detail of our sad predicament. Carefully, I eased my legs over the gunwale and groped for a foothold on the rocky river bottom. The next flash revealed Lawrence waist-deep in the torrent, struggling to release the canoe. Together we pulled it free and edged it towards the shore. With a heave we turned it over, lifted the bow onto the bank, and propped the stern up on deadfall.

For several hours we sat under the canoe, shivering in our wet clothes, our heads bent forward to avoid the thwarts. When the storm petered out, we paddled across the river and located the portage. The extra weight of our water-logged gear forced us to make two trips, so it was nearly dawn before we were on our way again. Lawrence remembered a deserted cabin some miles down river and we decided to make this our objective.

At daybreak, the birds began their morning concert. The first movement of the woodland symphony opened with the clear notes of white-throated sparrows. The fluted notes of thrushes picked up the melody. As if on cue from an unseen conductor, the rapid rhythms of yellow-throats harmonized with the performance and, from the hardwood ridge, rose-breasted grosbeaks introduced a quickening beat. Suddenly, the flute section dominated the piece, as wood thrushes, hermits, and veeries made their concerto-like entrances. Downy woodpeckers, the tympanists, pounded vigorously on dead trees as if to make up for their unmelodious song. By sunup, the valley was reverberating with music.

With a flash of crimson, a pileated woodpecker landed on a dead tree just ahead of us, announcing his arrival with the shrill squawk of a maladjusted public address system. Lumbermen once considered these richly coloured crow-sized birds to be their enemies, criminals to be shot on sight in retaliation for the deep rectangular holes they bored in trees. The fact that woodpeckers bored holes in trees already riddled by carpenter ants and therefore of little value as timber was completely overlooked by the firing squads. When it comes to dealing with his environment,

M.J.

man is a slow learner. Our own generation, ignoring past mistakes, seems to repeat a different absurdity: 'Over-mature trees are no good; cut them down; haul them away; make room for stands of trees of the same age.' The pileated woodpecker may yet disappear, a victim of man's excessive management of nature.

A ruffed grouse whirred into a nearby spruce. Lawrence aimed his paddle and in dumb show brought the bird down. Moments later we heard him drumming, probably poised on a mossy log, tail fanned, wings cupped. Indians used to think grouse drummed by hammering their wings against a log; that's why they called the ruffed grouse the 'Carpenter Bird'. Wiser in some ways, we know that the bird's only carpentry is the trimming of buds and twigs off deciduous trees during the long winter, when food is particularly hard to find.

It was five-thirty when we arrived at the cabin. Behind the long-deserted building was a clearing, patterned with ruins of what had been a flourishing lumber camp fifty years earlier. How quickly nature rubs out man's efforts! Here, a tumble-down camboose (from the French *cambuse*, meaning storeroom), aspen trees replacing its vanished roof; there a jungle of raspberry canes thatching a shored-up hole that had once been a root cellar. Only the cabin, built after the camp had been abandoned, could be recognized as a place of human habitation — cheating time for another five or ten years.

The life of the roof is the life of a cabin. Always the first to break down, the roof is the weakest link. First a small leak appears — a flaw through which the rain seeps. Inside, spreading brown rot indicates a vulnerable plank. One day the weakened board breaks, and the structure is doomed. Another break, then another. The wound is bone-deep and, skeleton-like, the beams are exposed. Year after year, in rain and snow, the roof decays and finally the skeleton collapses. Death comes quickly now. Floor boards powder; walls without roof-ties sag and buckle. Leaves and pine needles enrich the rotting floor-planks, and vegetation gains a foothold in a perfect substrate. Devil's club fills in yesterday's window openings, and balsam fir saplings line decaying door sills. The cabin is no more.

This cabin was in the early stages of disintegration. In a back corner, a broken board had left a hole in the roof large enough to draw the smoke from Lawrence's cigar straight upwards. A stove stood in one corner, but its pipes had rusted out long ago. The stovepipe hole in the roof stood open to the sky. Some previous canoeists, in return perhaps for a few days' shelter, had applied first aid to the ailing building: a piece of tin nailed over a hole in the roof, a broken window mended with a strip of canvas, a lard pail nailed to a rafter under a leaky board — minor treatment when major surgery was required. Under the lard pail the floor was rotting. The door, misshapen by the sagging lintel, would only close part-way.

We shared our accommodation with two pairs of barn swallows, which adroitly zoomed in and out the partially open doorway. These more established tenants made it clear that they resented our obstructing their aerial pathway, so we obligingly moved our packs to one side. It wasn't long before we met other guests in this Main Street hotel: the mice that frolicked across the floor in anticipation of the free lunches we humans would provide. This species, which biologists call *Peromyscus*, is more attractive than its urban relative, the house mouse. Their town-bred cousins are all grey, matching the unpleasant pallor of our cities and giving the impression of being unclean. *Peromyscus*, too, are colourmatched to their environment — but to the fields and woods in which they make their homes. Appropriately they are white on the underside with shades of tree-trunk brown on their backs. Only a taxonomist could inflict the unflattering name 'mice' on these attractive woodland creatures. They do, however, share a taste for many of the same foods that man himself enjoys. We thwarted their nocturnal rummaging in our food pack by suspending our supplies from a roof beam with a piece of wire.

It was afternoon when I awoke. Lawrence had gone fishing, despite the drizzle and the unpleasant low mist which hung in the trees. By the time he returned, I had a fire blazing in the old stove and the cabin filled with smoke. He came through the doorway with three trout, each about twelve inches long, dangling from a forked stick. 'Wasn't thinking of smoking them,' he remarked drily. 'Just thought we might fry them up.'

There are two ways to cook trout, each claimed by its advocates to be the only proper way. The first is to swab the pan with lots of bacon grease, get it sizzling hot, then pop the fish into it. This way, the 'hot pan' disciples claim, reduces the loss of moisture and makes the flesh brown and tender. The other method, greatly vaunted by the 'slow and easy' school of woodland cookery, is to fry the trout in a pan that is not too hot, over a slow fire, so the fish will not shrivel up and break apart. For generations, over countless campfires, the pros and cons of these two techniques have been loudly argued, providing a pleasant controversy which will go on as long as trout are to be caught and men and women hanker after wilderness living. Lawrence and I settled on the hot pan technique without much discussion — it was faster and we were both hungry.

For three days the rain continued, and, since wolves rarely howl in such weather, we holed up in the cabin. In Lawrence's years of experience, he had seldom heard of wolves being caught by trappers during rainstorms. This observation led me to formulate the hypothesis (which later experiences confirmed) that wolf howling is associated with wolf travel, when the members of the pack are separated from each other; howling is a means of keeping in touch while out on the hunt. It was Lawrence's observation, and ones similar to it, that led to my subsequent research on the subject of wolf communication.

The weather continued to alternate between drizzle and pelting rain. During clearer periods we fished and explored, but the fishing, after our first success, was a disappointment. Lawrence had caught his trout in a deep pool upstream from the cabin, where a spring entered the river. Water, in rivers like the Tim, warms quickly in summer, and trout gather in schools where springs empty into the mainstream. When the water is high, as it was at this time, the general temperature of the river is low and the trout are still dispersed. This, we reasoned, was the explanation of our failure to catch more fish. At least it provided us with an excuse, and what fisherman doesn't need an excuse now and then?

We explored the old logging camp with the eagerness of archeologists on the trail of history. Grass grows long over these sites and frequently conceals such relics as crosscut saws, axes, peevies, bits of harness, old quart whisky bottles, sometimes an old boot worn through at the toe, fragments of suspenders — odds and ends — souvenirs of shantymen and shanty days. But this camp was close to the river highway, and any treasures it once held had already been searched for and removed. We only managed to get torn by the tangle of raspberry canes and drenched by the rain.

It was three days before the sky cleared and when it did, late in the afternoon of the third day, we set out again. We planned to return at dawn after a night's howling, then, the following day, to leave the Tim by a portage near the cabin.

A few miles down river our route wound through another marsh where, here and there, small channels joined the mainstream. These tributaries marked their junction with the river by streaking it with translucent bars of clearer water, but gradually lost their identities as the bars vanished in the muddy waters of the Tim. At what appeared to be regular distances, redwing blackbirds, swaying on spindly rushes, loudly proclaimed their individual territories. In a few weeks, when their young would be strong enough to fly, they would move south to the orchard lands and enjoy the fruits of civilization. Acetylene exploders, oral sterilants, high pressure hoses, shotguns — nothing would discourage them: man had laid out a feast and to refuse it would be to renounce the very adaptability to which the species owed its flourishing existence.

Pond lilies slapped against our bow, arrowhead and bur-reed pulled at our keel. The roots of floating water-arum, cut through by muskrat teeth, bumped against our sides. Conditions on the water-turnpike were slow going for canoes.

'Dig in,' urged Lawrence, 'or the redwings will be out of here before we are.'

This marsh was a muskrat slum, unkempt, crowded, and untidy. The river banks were crisscrossed with a maze of muddy trails that must have confused the animals themselves. A muffled splash, then telltale ripples alongside the canoe, marked the spot where a rat dived out of our way. Overcrowding is the forerunner of a serious drop-off in numbers. Disease takes over, as the close contact between individuals causes it to spread and to claim more and more victims. But the effects of overpopulation can be more subtle: living at such close quarters the competition for food and places to live becomes intense, and the animals become intolerant of each other. This stress of crowded living in a complex social environment affects reproduction, and numbers decrease. Nobody knows if muskrats experience other, less evident effects of overcrowding: soaring crime rate, pollution, reduced life span, economic imbalance. If they do, they are unaware of the real cause of their miseries — a rising birthrate that will eventually annihilate their society. Doubtless they would be incensed at the idea of some governing muskrat interfering with their right to have as many babies as they want. They probably care little if the constant comings and goings of their offspring beat out trails that infringe on the grounds of others, or if their carefree gorging on water-arum causes their neighbours to go hungry. The scientific muskrat will pull them through, will discover a process that converts muskrat dung into high-quality protein, and invent high-rise burrows and twelve-lane supertrails. Muskrat technology is so advanced that only a heretic among them would call muskrats 'animals' or suggest that the same set of biological rules govern muskrats as govern lower forms of life, like frogs and snakes.

Four years later I visited this place again. The trails were gone. No muskrats dived alongside my canoe. The marsh was quiet.

'Black sky in the northwest — we're in for another bath.'

By now I had learned to heed Lawrence's simple weather lore. Five minutes later, the storm let go in a downpour. Marsh grass bent submissively before a pelting rain. If we were to reach our camp before dark, there would be stiff paddling ahead.

It was still raining when we landed and turned the canoe over onto its gunwales, then trudged up the hill to the cabin. We coaxed a reluctant fire from the old stove and blinked away tears from our smoke-filled eyes. Sardines, the consolation supper of unlucky fishermen, seemed almost as tasty as trout to our sharpened appetites. As I lay back in my sleeping bag, enjoying the best pipe of the day, I spotted a new leak in the roof. Shifting my bed a few feet out of range, I reflected upon what might have been the history of this lonesome old cabin.

It was evident that the lumber camp had ceased to operate when the cabin was built. A forest ranger probably received orders from headquarters one day, saying, 'We need a cabin part-way down the Tim. Had a fire in that back country last year, so we need a man there. Also heard that illegal fur is being taken out in the winter. You go — take a man with you. Build a cabin. Get it built by fall.'

The rangers would have pulled in at the head of the rapids, just as we did, then tramped around to find a good spot for their headquarters.

'Here's a good place. High, overlooking the rapids. Ground's firm. Lots of good pine handy — 'bout the right size, too.'

You could see by the neat way the tree trunks had been notched that the builders had taken pride in their work. The ends of each log were hewn to match the ones which would keep them company for years to come. This was no prefabricated cabin, but a shelter created from forest trees with a few simple tools in the hands of skilled workmen, who knew that the sounder they built it the more it would repay them in warmth and comfort during the long winter. I speculated on how, in times past, the builders had looked up at these walls and this roof and felt a sense of accomplishment in the straight line of notches and the snug fit of the ridgepole.

The floorboards had probably been salvaged from the shantymen's buildings in the lumber camp. But how were they brought to the lumber camp in the first place? There were no sawmills within miles. And the stove? Far too heavy to bring in by canoe, so how did it get there? At one time there must have been another road besides the river turnpike over which we travelled. But in all our explorations we could find no trace of where it had been. Time, one of nature's magicians, had caused the old road to disappear. We concluded that, instead of travelling on the river, the lumbermen had bumped in from the east in a wagon behind a team of horses.

The rangers, these second-generation pioneers, would have faced winters of lonely isolation. At the end of each day's patrol, they would ram their snowshoes into a drift and push open the cabin door. Soon the glass in the windows would steam up and the lamplight reflect on every pane. The top of the old stove would grow cherry red as the chill retreated against the well-chinked walls. Outside, trees made brittle by the frost would sound their wintry rifle fire, and the roar of the rapids would be silenced behind half-drawn curtains of ice.

Then, years later, would come the word from headquarters: 'No need to put a patrol on the Tim this year. Aircraft from Pembroke can do the job for less money.' That first winter of abandonment must have been a shock to the cabin. No fire leaped in the ancient stove; no light shone out from the coal-oil lamp on the table. Wind, pushing open the unlatched door, would spread a crescent-shaped snowdrift across the floor. In the spring, the boards would swell and warp. The door would never completely close again.

But the cabin's life was not over. Summer after summer it would shelter men and women whose love for the wilderness would take them down the Tim. If it were a new structure, it would have no place in a wilderness recreation area like Algonquin. But the old cabin was much more than merely a shelter; it was a relic of the early days, and a reminder of the first men who entered the wilderness. It relates man to his past and to the processes of change in nature, in technology, in values.

The next day Lawrence and I left the Tim. The rain had stopped, but clouds still hung low. It was hard work paddling against the current; the river had risen a few inches during the night. We were returning to the Research Station with no trophy; that meant we must paddle the Tim again, so secretly we were glad. But we saw a big wolf track, made within twenty-four hours, on a muddy bank a mile above the cabin. Was it a single wolf, out travelling despite the rain? We howled, but bird song drowned us out.

As we shouldered our packs, I looked back over the river. It had become a partner in our adventure. We had floundered waist-deep in it, dragged our lines through it, sliced it with the bow of our canoe, and sent ripples lapping at its banks. At a gentler season we would travel this way again, when the spring freshet would have spent itself, the nights would be warm, and moonlight would flood the valley.

4
Night Trail

Hazel brush wet with dew. Bracken ferns hiding uneven ground. Spindly limbs of fallen trees reaching out wherever you shine your flashlight. Cool air in the lowlands. Stars sparkling above. Feeble shadows and dim outlines. Moonlight and aurora borealis. This is the night trail.

Man is a creature of the day for a simple biological reason. The pupil of his eye is made up of two types of cells: *rods* and *cones.* Only rods are sensitive to low light intensities, and of these, man's eye has few compared with night creatures. Our most important form of protection, sight, diminishes as the sun goes down. Danger, whether it be a mugger on a city's back street or a rock jutting out in front of the canoe, appears seconds before it is too late as a dim uncertain shadow. Realizing this handicap, man shuns the darkness.

But in the north woods, most mammals, such as otter, beaver, fox, and wolf, are primarily creatures of the night. To penetrate their world we must tolerate the cold dew that plasters our pant legs to our bruised shins as we push through the undergrowth, and relish the grip of the paddle in our numbed hands as our canoe cuts through the damp air hanging over the marsh. We must learn the phases of the moon, the time it rises and sets, the direction the Big Dipper swings around the North Star, the nightly path of Mars, and the cold light of Jupiter. We must learn to distinguish the night sounds — tree frog from wood frog, barred owl from horned owl, and the loon, whip-poor-will, and bittern. We must decipher the difference between the leaf rustlings of white-footed mice and those of woodland jumping mice, between fox and snowshoe hare, and the frightened crashing of a moose from that of a bear. Remarkably, when we learn these things, the wet legs and numb hands are forgotten. The night trail becomes a wilderness experience of great value, worth seeking out.

Lawrence and I had no choice. For us it had to be the night trail. During the spring, loggers had reported hearing wolves at Hiram Lake, about five miles north of the highway. Although they had seen only one wolf loping along a logging road, they had often noticed wolf tracks embossing the soft, newly bulldozed earth. Sometimes wolves had howled in reply to the shattering clamour of their chain saws. Were the howls a protest, a canine curse at man's arrogance in destroying the sweet sounds of the wilderness?

At Sunday Lake, only four miles from Hiram, fishermen had frequently reported hearing wolves. Were these the same wolves? The accuracy of our census depended upon answering such questions. And to find the answer we had no choice but to follow the night trail — or, as it turned out, two trails. Lawrence would hike a fisherman's trail on the east side of Sunday Creek; I would walk an abandoned logging road as far as possible, then follow game trails to Hiram Lake.

We parted at the highway at three o'clock in the afternoon, leaving me enough time to reach Hiram before dark. Lawrence called out words of advice as I walked away: 'Watch for small things. A tree stump torn open by a bear, the direction the ferns is bent along the game trails. Maybe you won't remember all them things when you start back, but it'll come to you — sort of unrolls as you go. Seeing one thing'll remind you of the next.'

The sun was hot and I was perspiring. There was no wind, and a few early deer flies persisted in making a reconnaissance around my head, looking for a meal of blood. I had forgotten my cap, and thought it better to keep batting at them before they burrowed into my hair and raised large welts. I had no trouble following the trail the first half mile. It led to a campground dump. No bears were in sight — too hot to search for food, but wait until night.

Many tourists follow their noses to the dump, even though the road is marked 'closed'. Seeing bears in such degrading surroundings is a real 'attraction' in many North American parks. Bear-proof garbage cans, steel holding bins, and highways leading to city dumps not too far away provide sensible alternatives, but they cost money. So just dump the garbage out of sight, and shoot the bears if it attracts them to nearby campsites. In parks like Algonquin, with a density of one or two bears to a square mile, such mortality is insignificant to the total population. But what kind of a philosophy is that? Don't individual animals have rights in parks?

As soon as I had hiked north of the dump, its stench disappeared. The trail narrowed and I remembered Lawrence's advice. In a wet spot ahead, two logs were lying across the mud; their bark had gone, and I knew they'd probably be slippery. A good place to check for wolf tracks, but there were no signs of any. A bear had walked in the soft earth, but that was all.

'Gottcha, deer fly.' After that, only two of them were buzzing me.

A ruffed grouse, straight ahead on the trail, was having a dust bath, flailing the sand over itself. That's a curious habit, and part of the preening process; a way to arrange its contour feathers neatly and hook the tiny barbules along each feather's edge to those adjacent. The ruffed grouse is a diurnal creature, like man, and roosts in the spruces at night out of reach of fox and wolf. An owl could get it, though, so it's never really safe. Its security as a species is assured by a large clutch size and five or six young to each pair every year. A quick step forward and it was into the air in a whir of feathers, leaving a very noticeable sandy depression that I resolved to find with my light on the way back.

Bluebird Lake. Not much of a lake — just a series of two or three interconnected beaver ponds stretching eastward, each one at a slightly different elevation. I sat down on a grassy knoll to scan the ponds.

The warm sun beat down on my neck. I lay back and recalled my trail markers: the grouse dust bath, the slippery logs, and the bear's tracks. Lawrence's advice could be boiled down into one word — perception. *See* your surroundings, don't just pass through them with your mind on other things. Come to think of it, much of life that passes close to a person in the course of a day never reaches the conscious level of his brain. How many objects that actually impinge on the eye do we really see? How many sounds do we really hear, or odours do we really smell? How many of our actions, through repetition, have become reflex?

Few city dwellers hear traffic rushing by, or aircraft overhead, or fire sirens, or the clack of heels in the room above. They only see rows of parked cars when they happen to be searching for that last vacant place. Rarely do they perceive expressions on the faces that pass them on the street, or know the number of steps to their veranda that they walk up and down each day. Perhaps this is good: if we allowed our minds to be harassed by every urban distraction, how could we accomplish anything? The only way to find peace in the city may be to build a box around our minds and connect an automatic switch to the sensory neurons entering our brains, thus sidetracking irrelevant impressions before they bother our consciousness; in other words, to shut out the unessential so that only the cry of FIRE shouted in our ears makes us aware of our surroundings.

But it's not so easy as that. The circuit must either remain connected and feed in everything, or be broken and, like high-voltage potential across a spark plug, react only to those sensory impressions which are strong enough to jump the gap. A man whose mind has the circuit broken will never appreciate the stars in the city sky, or the orange flash of orioles nesting in elm branches hanging over the street, or the merry call of chickadees in the ravine outside the office window. He will not realize that the sky has become cloudy until it begins to rain, and only then will the city man seek shelter — fortunately never far away. But what of the man who lives in the country, whose daily work takes him far from shelter? Whether he is ploughing the 'back forty' or chopping down a tree two miles from camp, for him clouds have their special meaning. He does not wait until the storm hits to find shelter. A warm April breeze, meaningless to the city dweller, whispers change to the farmer, promises a good start for his winter wheat as it melts away the lingering patches of snow along the fence lines. A man living beside a fast-flowing river is seldom conscious of the sound of rushing water, but immediately aware of any change in its cadence. Let a log get caught crosswise to the flow and he is instantly alert. To him the river is important. It may be the source of water for his livestock, or irrigation for his fields. Its grassy banks may provide a set for his mink traps. To the city man, the river of cars in front of his office is only significant at five o'clock when it is time to go home.

A splash. I sat up. V-shaped ripples fanned out across the pond. The angular head of a beaver surfaced among the water lilies, so close that I could almost reach out and touch it. Neither of us moved. Finally it decided that I must be inanimate, and began feeding, pulling up the juicy underwater roots. It was time for me to get along — across the beaver dam, through the alders, up the hill, and into the ferns again. Good area for beaver with the ponds and all these aspens on the high ground. To a beaver colony, abundance of aspen means large litters of kits and longer occupancy of their pond. And a good supply of beaver means more food for wolves, at least in the summer. So wolves and aspen trees are linked together in one of the many indirect chains that bind living ecosystems into a complex unity.

A red-tailed hawk screamed overhead. I stared up into the sky as it circled and glided. But I was watching the hawk, not my surroundings; how easy it is to climb a hill without really seeing it. Suddenly, right in the middle of the trail, was a wolf track. One step more and I would have walked on it. There must have been a seepage of water, as the sand was wet. Too bad I hadn't brought along my bag of plaster; that paw print would have made a fine cast. All of four inches long. Strange that a wolf's hind foot is so poorly adapted to running. Although he is up on his toes, the next series of bones (corresponding to the sole of our feet) angles obliquely up and backwards. If he had been perfectly adapted for running, these bones would come straight up at right angles to the toes and pitch the animal forward, as they do in the cat family. This seemingly insignificant feature of evolution has had far-reaching effects. It has left the wolf at a disadvantage, since most prey animals, provided they are healthy, are faster. Thus wolves are normally forced to use group tactics to capture large prey.

The need for wolves to form packs may to some degree be the result of an incomplete right angle between the bones of the feet.

This was the track of only one wolf — a lone hunter. The wolf had crossed at right angles to the trail; strange he hadn't turned and followed it. If he had continued his course in a direct line for four miles it would have taken him to Sunday Lake. Would Lawrence have heard him?

Overhead, a myrtle warbler was singing in the maple branches. I walked faster now — downhill. The trail swung out of the trees and into a marsh shaped like a bow tie, crossing the marsh at the centre of the bow on an old bridge that formed the knot. Either the water level of the marsh had risen, or the pilings of the bridge had sunk, for it was covered with about four inches of water. That wasn't so bad, except that the bridge had a great deal of spring to it; the centre pilings must have disappeared completely into the water. I edged my way along the slippery logs, trying to make myself light by imagining an invisible hand lifting me up by the shirt collar. The water came up and up — two inches from the top of my boot, one inch, then two inches again and I was across. I wondered if every load of logs that crossed the bridge

made it safely. I pictured wheels mired in mud, horses splashing, men cursing.

Seven-thirty. The sun was much lower than when I was lying on the knoll at Bluebird Lake, and its colour had changed to orange. Over on the far side of the marsh it was creating shadows in the branches of the balsams and spruce trees, exaggerating the three-dimensional effect.

Up into the hardwoods, where the deer flies found me again. I tried running as far as a bend in the trail, but they stayed with me. Always three flies — that must be the fly-carrying capacity of the cubic foot of air around my head.

I wondered how Lawrence was doing at Sunday Lake. He had taken his fishing rod. Right at this minute he was probably casting his line off some rocky point. I hoped he would catch something; lake trout would make a good supper for us.

I made a mental note to remember a wet spot on the way out. Best if I were to leave the trail on the west side and circle along the base of the hill. Beavers had built a new dam, bringing the water up into the trees.

Alongside Blackfox Lake and an old lumber camp; the trail led me out into a clearing of crumpling buildings. The forest had already marched back into the clearing and considerably reduced its size. A huge ash tree had fallen, and laced through its decaying branches were raspberry bushes that scratched my arms as I climbed over it. I sat down on a stump by the water and watched the tiny wavelets raise and lower some duckweed floating between two sticks. There is something sad about the abandoned artifacts of man. Is it an uneasy feeling that the best is gone, that life has passed by, that somehow the old days were really better? Or is it a submerged fear that the lumber camp is an illustration of man's careless disregard for his natural heritage, the first stage of a blight which may engulf our civilization? When all the merchantable wood had been cut, there was no money to pay the food bill. The capacity of this small area to support humans by trading one essential resource (wood) for another (food) was cut to nil, just as the capacity of larger areas may some day be cut because they can produce no more oil, or no more clean water or air. In the 1930s, when this camp was abandoned, the loggers merely moved to uncut timber land. Migration was the antidote for dwindling resources, and the practice was to cut, burn, and move on. But today we don't talk of resource exhaustion in terms of lumber camps: now the boundaries of scarcity are counties, provinces, states, and, in many parts of the world, whole countries.

Deep shadows hung over the water. The sun had left the clearing, and coolness was settling in around the old camp. I lined up my compass on the map and calculated north five degrees east to the foot of the lake.

'Cut deep when you blaze,' Lawrence had said. 'If you just take the bark off, the blazes turn brown and you can't pick them out with your flashlight.' Now and then, through the maple underbrush, I caught a glimpse of Blackfox Lake. At the north end of the lake a stream had cut a deep gully into the gravelly glacial till, and I scooped up a drink at its mossy edge. I told myself that on the way back, if I were too far north to hit the top of the lake, I'd know by this gully when I'd come far enough. As a precaution, I made a quick sketch in my notebook.

I swung almost due east and into a growth of tall hemlocks. The trunks were riddled with sapsucker cuttings. Such methodical wood carvers: every hole was in perfect alignment! After I left the hemlock stand, I was again in hardwoods. A deep mat of decaying leaves gave spring to my step, and soon I arrived at Redfox Creek, where a few inches of water meandered between stones that once formed a meltwater channel from the great glaciers.

There was a deer fifty feet upstream. Even in the half-light I could see that he was a buck, his antlers budded with velvety growth knobs. He hadn't seen me, so I tested his reaction to a wolf howl. At the sound he looked up quickly, walked a few paces, then lowered his head and resumed his drinking. I howled once more. Again he looked up, this time his tail twitching nervously. He took a few steps and disappeared into the trees. His slowness to recognize danger might some day cost him his life. However, if I were a wolf, I would not give myself away with a howl. The fact that deer do not always run when they hear wolves argues that wolves remain silent when they hunt.

It was getting dark when Hiram Lake appeared as white specks between the dark spruces. I should have made it earlier so I could have climbed a hill to howl from while it was still daylight. Instead I decided to try a north bearing and keep on it to a ridge top. It was quite dark by that time, and I switched on my light.

'Make sure you don't hold your compass too near your light when you read it,' I remembered Lawrence saying. I knew the metal would deflect the needle if I did.

The first few hundred yards were scrambly — up and over fallen spruces. First I was in alder, and the ground was wet. A loud splash over to the left brought me to a sudden stop. Only a beaver slapping its tail, so I kept going. The ground began to rise and the alder gave way to a jungle of small aspens. A whir at my feet, and a woodcock slanted upwards across the beam of my light. Although it was late for a breeding display, I waited to hear his nuptial flight. There it was, the whistle of wind rushing through wing feathers. Faster and faster came the bursts of whistling. A pause, then the brief twittering song at the summit of the spiralling. Silence as the bird plunged with wings half folded. 'Beep'. It was back on the ground.

At night, man's hearing makes some compensation for his diminished vision, but never enough to give him the assurance he has by day. With darkness comes an uneasiness, especially when we are in an unfamiliar place. We fight against darkness, draping our city streets with lights, closing our curtains in the evening to shut out that black rectangle of unknown danger. The folklore of the pioneers is full of references to the hostility of the night forest, lying threateningly out there beyond the edge of the clearing.

Deer hair. My light fell on the scene of a wild drama that had taken place one winter's night. Scattered on the moss were the remains of a deer — a leg bone picked clean, a lower jaw, and mounds of hair. Everything else was gone. I picked up the jawbone. The deer had been old, approaching its ninth year, with the ridges and crests of its molar teeth flattened by wear. I crammed the jaw into my jacket pocket to add to my collection. The data we were gathering would tell us more about the impact of wolves on deer.

As it turned out, this jaw was included in results published a few years later, which showed that during the winters of the study twenty-two per cent of the deer killed by wolves were eight years old or older. This was a higher figure than this age group represented in the live deer population, establishing that wolves are more successful at killing older animals than those in the prime of life. This was no insignificant discovery. Combined with observations that wolves are not always successful when they hunt deer, even when they use stealth, it led to an interesting but difficult question: are wolves culling the weak, the diseased, the young, and the old out of the herd, leaving the deer population healthier and more vigorous? The conclusion that they are was made from other studies: on Isle Royale in Lake Superior where wolves hunt moose, and in Alaska where, besides moose, they hunt caribou and Dall sheep. The answers to such questions, as to so many questions about living things, have clever ways of posing new questions. Would those weaker animals that fall victim to wolves die anyway from natural causes, such as disease or weather, before the next breeding season? If so, wolves are merely over-eager scavengers, having no real influence on the quality of deer herds. But if wolves cull out inferior animals that otherwise have no quality beyond the ability to hang on to life and to produce inferior offspring, their influence in terms of pruning the genetic hedge may be great. How do we find out? If we were studying single-celled organisms in a dish of water, one valid approach sometimes used to discover principles of biology, we could design an experiment to test the interactions between predator and prey. But the answers we received to our questions might not stand up outside the laboratory. Deer and wolves may have their own special relationships, complicated many times by a host of interlocking factors such as the abundance of other prey species, the food preferences of wolves, and even the density of the forest cover.

My deer had made his last leap over a large spruce log. Before he leaped, canine teeth had sunk into his flank, pulling out a wad of hair that still lay beside the log. Blind with fright, he had failed to see a second wolf who, anticipating the deer's line of travel, waited where I was standing. It was all over quickly — one wolf on his neck and one on his underside. And so the ancient drama was played, the drama that early evolutionists claimed was the 'stuff' of evolution: Tennyson's 'nature red in tooth and claw'. Today's scientists regard predation as only one of many mechanisms that make the evolutionary clock tick. But it is still significant: only the fittest must survive to leave offspring, or living populations will degenerate into genetic rubbish.

The night was growing colder. I zipped up my jacket and plunged on, pausing now and then to blaze a tree. The ground continued to rise and I was in hardwoods, then out into an opening. A whip-poor-will, singing his roller-coaster song somewhere below me, was suddenly joined by another, whose song was higher pitched and faster. They stopped. Bullfrogs chorused rhythmically as though taking their time from the baton of an unseen leader. The whip-poor-wills began again, upstaging the lily-pad section. Abruptly the chorus stopped, and the silence pressed down, closed in, restrained every blade of grass, every maple leaf in its mute grasp. Suddenly a white-throat broke free, but its notes were muffled by the thick vegetation. Off towards Hiram Lake a bittern began pumping.

I howled, and my voice seemed to float away into the darkness. Nothing answered; the bittern, undisturbed, continued pumping.

Through the trees in the southwest I could see the moon, a few nights past being full. The hills below me took shape, stretching out like features on a gigantic aerial photograph. They looked indestructible: here since the Ice Age, here for thousands of years to come. Left alone, the throbbing life of the wilderness could go on without end. The nitrogen, phosphorus, and calcium locked in the tree trunks for the past hundred years will return to the soil, become altered by countless decay bacteria, and cycle again through living things. On their next cycle they may spring open the seed coat of a bunchberry, or be drawn once again into the trunk of a maple to form a cell wall of wood fibre or a resin that lodges in the heartwood. All this will happen if the wilderness is left alone. But even if no man were to set foot on this hill again, these delicate pathways can still be destroyed. Setting aside parks and wilderness areas may give us assurance that some nitrogen atoms will never be locked underground by cement, but it will not guarantee their presence in green leaves. Pulp mill wastes, smoke from smelters, or radioactive fallout may prevent that.

Pesticides, carried miles by the rain, may alter bird and insect populations, leaving the forest open to disease. The contrails from thousands of jet aircraft may even reduce the permeability of our atmosphere to solar radiation. Then the nitrogen would never again be part of an amino acid in the protoplasm of a birch leaf waving in the summer wind. The ecological boundaries of a wilderness cannot be drawn on a map. They are the whole world: land, water, and the enveloping atmosphere.

I started back, downhill, due south. My light caught the blazes I had made on the trees, then the deer hair by the spruce log. My afternoon journey was rolling back, just as Lawrence had said it would. The whip-poor-will was calling again, this time on my right, exactly where I expected him. I continued down into the alders where the woodcock flew, then into the spruce deadfall, and soon I saw the outline of Hiram Lake. I sat down on a fallen log that extended out into the water. The air felt cool through my damp shirt. A trout splashed on the surface and I wished I had my fishing rod. As I lit my pipe I pictured Lawrence, over on Sunday Lake, sitting in a similar grove of spruces, puffing on his cigar. A few mosquitoes found me, and I snapped on my light and watched one trying vainly to bite through my pant leg. Finally I ended its frustration with a quick slap.

I shut off the flashlight and gazed up at the Big Dipper. Go five times the distance between the front two stars in the direction they point, and there's the North Star. For centuries, on land and sea, travellers have looked to that reassuring pinpoint of light. I wonder how early astronomers discovered that this one star in all heaven's galaxies never moves in relation to the earth?

I howled six times, pausing between howls only long enough to suck in a breath of cool night air. I pictured the sound drifting across the lake and into the spruces on the far shore. It rolled on, growing thinner as it pushed through openings in the hills, thinner still as it moved along hardwood ridges. Finally, exhausted, it spent itself in a rustle of birch leaves a mile or so away. As it rolled along, snowshoe hares suspended their games of tag in the moonlight clearings, instantly alert; bullfrogs stopped their chorusing. The nerves of every forest creature grew tense: the wolf — the hunter — was abroad.

I got up and switched on my light. There, not a foot away, was one of the north woods' most remarkable predators. What a triumph for evolution to have fashioned a meat eater with such cunning that throughout its life it remains in one place, waiting for its prey to come. Not a bear, or a mink, but a pitcher plant. The leaves, shaped like hollow tubes, collect the rain and the dew to form a living insect trap. Stiff hairs, pointing downwards, prohibit escape; the executioner drowns its victims. Sometimes a trapped mosquito, attracted by the stagnant water, blindly lays its eggs within the prison, unaware that its offspring will never escape. Enzymes produced by plant cells dissolve the unfortunate victims. Is it more cruel to drown and dissolve a victim than to rip him apart with wolf teeth? Or is cruelty only a subjective concept held by man?

Back along Redfox Creek, back through the hardwoods with the springy leaf mat, back through the hemlocks — it all unfolded as it should. I stopped at the same little gully to drink and check my direction — south five degrees west — the converse of my bearing on the way in. Up through the hardwoods, scrambling over deadfall that in daylight I had been able to walk around. The old lumber camp. I shone my light through a dark doorway. It reflected on a shattered windowpane.

I sat down on the stump at the water's edge. The duckweed was perfectly still, now. A low mist hung over the lake, caused by an air inversion. In the deep woods air inversions at night are merely interesting phenomena; over cities they are cause for anxiety. Normally, during the day, the sun heats up the air that is in contact with the surface of the earth, and the warmed air rises. When the sun is not shining, the lower air is cool and stays close to the ground. The cool air holds less moisture, and beads of mist are held suspended over Blackfox Lake. All it means here is tough navigation in a canoe and wet alders along the shore. But in a big city it means trapped air pollutants; it means dangerously low visibility for aircraft, and early morning traffic accidents on the freeways. When it continues night and day, it means bronchitis, asthma, emphysema — and death.

I howled again. Far away to the east an answer floated back. It sounded like a single wolf. Suppose it was only Lawrence?

Wouldn't that be a joke, each of us travelling miles through the bush just to howl at each other. No, there was certainly more than one animal — a whole pack! Their notes rose and fell. Were they harmonizing, simply howling for enjoyment, or were they trying to communicate with other wolves at some distant point — absent members of their own pack or those of another? Perhaps they were reaffirming their social positions within the pack. Possibly all these explanations provide the stimuli to howl. But, in the light of later observations, I doubt that the wolves were concerned with their social position, even though most social rituals in wolf packs serve to remind each animal of its rank. When we later made studies of the daylight behaviour of wild wolves, we confirmed that for the two or three minutes of the pack howl, all wolves are equal. Social subordinates, who are normally outcasts at breeding, peripheral in play, and secondary at times of food shortage, are just as prominent as social dominants when one wolf's impulse to howl triggers the whole pack. Perhaps this is because, unlike other social events, howling offers no direct threat to the status of the dominant wolf, and so he can tolerate this sudden outbreak of emotion by his inferiors.

I pulled out my compass to get an exact bearing. The pack should have been closer to Lawrence than to me. If he had heard it, we would have cross-bearings and could plot its exact location on the map. Fortunately, territories are small, or our job would have been impossible. Capable of having territories of one or two hundred square miles, the eastern forest wolves voluntarily restrict their hunting grounds all summer. When food is plentiful, why travel far? Pups must be considered. Even tundra wolves following the migratory barren-ground caribou must limit their travel while the pups are too young to follow. But within their territory, wolves move freely. For this pack, the distance from Sunday Lake, where the fishermen heard them, to Hiram Lake would be an easy one-night foray. They might even make it to Hiram and back during the hours of darkness.

On the trail again I felt surer of myself. There was the wet spot where beavers had raised the water level; I went around it to the west. The ferns were wet with dew, and my pants clung to my legs. Down to the bow-tie marsh and across the floating logs.

In my haste I dragged my foot and water slopped over the top of my boot. Another howl — no reply. Into the hardwoods again. I was making good time. There was the wolf track impressed in the sand. I pushed through the alders and now my shirt was soaked through. I crossed the foot of Bluebird Lake, and a beaver, perhaps the same one I had seen that afternoon, slapped its tail. I could smell the dump; its odour was drifting farther north in the dampness. Two eyes ahead: a bear on the trail. I stopped, my heart pounding. I knew the bear would run, so I kept walking. It tore away into the dump with a great clatter of flying tin cans. The highway was just ahead. I could smell cigar smoke.

'Any luck?' Lawrence's voice. The night trail was ended. We could add one more pack to the map at the Research Station.

What if we were truly creatures of the night, endowed with more rods in the pupils of our eyes? We might better understand the countless nightly dramas of the forest. With a chance to spread our activities over twenty-four hours, we could produce more — and waste more: the evening rush for campsites in our parks would be spread out — but we would lose the tranquillity of darkness, the gradual closing in of our perceptual field as daylight fades. With our perception diminished, perhaps we would think more.

And we would lose the humility of knowing that while man has many unique biological adaptations with which he has dominated the biotic world, other animals have a big advantage over us half the time — every night.

5
Wolf Legends

They were near; one wolf was less than two hundred feet away. No distant howl this time. We had unknowingly walked right into the pack. We switched off our lights and sat motionless. Howls still rang in our ears, minutes after the last chorus of yapping had abruptly stopped.

The old logging road we had been following ended at a partially dried-up beaver pond. The wolves were just in front of us, out there in the long grass. The night was clear and cold, and the moon, in its first quarter, had not yet risen.

They were spread out over a hundred yards or so. From near the centre of the focus of sound had come the short treble yipping of pups. It was too dark to see anything.

A twig snapped on our left. I looked over at Lawrence. His cigar had gone out. I felt like a carefully designed experiment, human flesh set out in a wolf rendezvous site to test by first-hand observation whether wolves attack people. Then I thought how similar the circumstances were to supposedly factual accounts by some pioneers of attacks by wolves. If these were true we should be climbing trees.

'Let's howl again,' whispered Lawrence. 'Maybe we can get a count of 'em.' Just as a precaution I estimated the distance to the nearest spruce as Lawrence gave a low howl. A wolf barked close on our left, probably the same animal that broke the twig. Then a bark and a howl straight out in the meadow. A puppy yip, and the whole chorus broke out. They had fanned out in the interval, and now formed a semicircle around us. The response was shorter, ending with a few deep barks from the wolf that had started it.

'Pretty hard to count,' Lawrence whispered. 'At least four adults and I don't know how many pups. Better give them more time before settin' them off again. Maybe they'll howl longer.'

I eased over off my cramped leg and settled more comfortably in the grass. When Lawrence didn't relight his cigar, I knew for sure he too was excited. Excited, but not afraid of being devoured. If all this had been happening a hundred years ago, and we were the first pioneers to enter this region, we might have been fearful. But we were viewing this wilderness spectacle with the knowledge that wolves consider man distasteful. In the light of modern biology, the bloodthirsty accounts of early pioneers are startling by contrast. Take, for example, the story of the young settler, Thomas Conant, as related by his grandson in *Life in Canada*. He recounts his grandfather's midnight walk home through the woods from his sweetheart's place in the early 1800s:

'Getting over about one-half the distance, he heard the distant baying of wolves. Fear would, it may be supposed, lend speed to his feet, but thinking rightly that he could not outstrip the wolf on foot, he walked quietly along, watching for a convenient tree for climbing. In a very few minutes the wolves were upon him, in full cry, eyes protruding, tongues lolling, and ready to devour him. A near-by beech tree, which his arms could encircle, furnished him with the means of escape. He climbed, and climbed, while the wolves surrounded him and watched his every motion, never ceasing their dismal howls the live-long night. Thus he kept his lonely vigil. To lose his hold for a single second meant instant death. Great, however, as was the tension upon his strained muscles, they held on. Morn tardily came at last, and with its first peep the wolves left him and were seen no more. When they were really gone, he said he for the first time began looking about him, and found, with all his climbing, he had ascended a very few feet from the ground, and but just out of the wolves' jaws as they made frantic jumps to reach him.'

What really happened? Nothing very different from what was happening to Lawrence and me. Thomas Conant came across a wolf rendezvous site; he interpreted the wolves' curiosity as attack, and up a tree he went. The wolves encircled him, as they did us, trying to scent the intruder. They stayed all night because, like us, he was in their rendezvous site. To have seen the protruding eyes and lolling tongues were remarkable feats of vision at midnight, considering that only when it was daylight did he see how far he had climbed up the tree.

'Did you hear that?' Lawrence broke into my analysis. 'A snarl. Sounds like they're fighting over something.'

Another snarl, a yip, then silence. Away off to the north a single wolf howled. The puppies out in the meadow let loose, and first one then another of the adult wolves howled with them. No barks this time, but the wolves were still halfway around us. Surely by then they knew that we were men, their dreaded enemies, the species who poisons them at their kills, strangles them with neck snares on their runways, and sets foothold traps around their dens that force them to chew their own living flesh in vain attempts to escape. Didn't they recognize us without rifle scopes to our eyes, or snowmobile helmets on our heads, or bulldozer blades in front of us? But they accepted our presence and went about their business: scrapping over a beaver carcass or the leg bone of a deer or whatever it was out there, and howling back to absent pack members.

Lawrence shifted into a half-crouched position. 'Let's give 'em another half-hour. See what that lone wolf does.'

Thomas Conant got off lightly. Poor old Corporal Crawford didn't fare so well. His fate, which will bring tears to your eyes, was laid down as fact by W. H. Higgins in *The Life and Times of Joseph Gould*, a pioneer of Upper Canada and later a member of Parliament. Crawford was an early settler, who built a cabin for his wife and three children less than a hundred miles from where Lawrence and I were sitting.

'That winter was one of unusual severity. Towards its close, the howling of the starving wolves was incessant throughout the night, and filled the poor woman and children with terror. Crawford had his gun, but only a scant supply of ammunition. He was in the habit of making short excursions, in order to get supplies of fresh venison, which he was always able to fetch to the hut before night to the expectant family. From one of these — the last in this tragic story — he did not return as usual. Night came on, and the uneasiness of the wife grew to alarm at the husband's absence. She "tooted the horn" again and again, but there was no answering response. Solitary wolves were seen prowling about the hut when the affrighted wife looked out at the door and "tooted" in the blinding snowstorm which came on. This it became unsafe for her to do any longer, so the door of the hut was kept shut and barred. Far into the night, the howling of the pack, at first distant, came nearer and nearer to the hut; the watching woman heard a rush past, and believing that her husband was pursued, in her fond eagerness to give him succour opened the door. Fatal step! The ferocious brutes rushed in, tumbling over each other in their bloody eagerness; the woman, with her infant in her arms, was knocked down, and the savage animals fought and tore each other in glutting on their defenceless prey. A child's crib, clumsily made of heavy timber, was overturned in the dreadful onslaught, and covered up in it was the little girl, the oldest of the children. She was completely hidden by the overturned crib, and, rendered unconscious by fright, never moved. Daylight broke in upon this horrible scene in the woods when Crawford arrived, to see the wolves, some of them with bloody jaws, slinking away from his wretched cabin. He had followed too far after a buck which he had wounded, and on his return home night had fallen, the snowstorm had come on, and pursued by a hungry pack which had got on the trail of the wounded deer, he had taken refuge in a tree. This was scarcely a mile distant from his home. While in this place of safety, the wolves, which were howling and jumping at the foot of the tree, suddenly followed in pursuit of the buck, which unluckily had taken a course that led towards the hut. Crawford could tell from the savage howling which arose above the storm that something terrible was going on in that direction. With the first streak of daylight he made his way for home, where, frantic and heartbroken in his agony, he encountered the tragic scene described.

'Turning up the crib, the little girl was found, unhurt. Some bloody tresses of his poor wife's hair, some fragments of clothing and the half-devoured carcass of a wolf were all that was left to shew the dreadful havoc that had taken place over the bodies of the defenceless woman and two younger children. Immediately outside the door of the hut were found the antlers of the deer, from which it was conjectured that the hunted animal ran towards the light when the door was opened by Mrs. Crawford.

'Corporal Crawford never again returned to civilized life. The events of that dreadful night completely unhinged his mind. Through his friendly Indian companion, he was adopted by the Mississaugas, who regarded his infirmity of mind as an additional reason for their protection. The little girl, whose life was so miraculously preserved from the ravening jaws of the wolves, was returned to relatives of her mother on the American side, and was afterwards lost sight of in the march of events and settlement.'

Did it really happen? We must take the word of the only witness, a man with an 'unhinged' mind. But what a perfect cover for murder! Kill his wife and child, bury them, blame it on wolves, and disappear with the Indians. Doubtless Crawford would have been unable to testify even if there had been an investigation or an inquiry, and it is doubtful whether one would have been held at that time.

'Nothing from our lone wolf. Let's try 'em again.'

This time, only one wolf answered. It was back at the far side of the meadow, about four hundred feet away. Maybe the novelty of entertaining humans was wearing off. The moon had risen, and a mist hung about fifteen feet above the half-dry meadow. Far away a loon called. Was it trying to get the wolves going, too?

Fear of wolves seemed to be a characteristic trait of pioneers. Explorers, on the other hand, many of whom also came from Europe in the early part of the last century, did not share their apprehension. Daniel Harmon, a partner in the North West Company, who travelled from Montreal most of the way to the Pacific coast, wrote, 'They [Wolves] are all very voracious; but they never attack a human being, unless when suffering greatly from hunger.' The last proviso leaves some room for fear. How can you tell if a wolf is hungry until it is too late?

Other explorers give wolves an even better billing. Captain Jonathan Carver, who travelled more than five thousand miles in the interior of North America in the late 1700s, wrote, ' . . . nor will they ever attack a man, except they have accidentally fed on those slain in battle.' About early days in the Canadian west, A. B. Stock, in *Ranching in the Canadian West*, commented in 1912, 'You need have no apprehension of danger to yourself from these grey timber-wolves, for I have never heard of one attacking a man.' Even in the Arctic, where it is reasonable to assume that wolves might sometimes have difficulty in obtaining food, early explorers seemed to show little fear of them. The first white man to cross the eastern Arctic, Samuel Hearne, wrote, 'All the wolves in Hudson's Bay are very shy of the human race, yet when sharp set, they frequently follow the Indians for several days, but always keep at a distance.' A later Arctic explorer, Vilhjalmur Stefansson, in his *Arctic Manual* (1944) written for the United States Army to aid airmen in trouble, stated, 'The wolf pack, as described in fiction and in news dispatches, does not exist . . . Interested in learning why such dispatches were carried, he [Stefansson] was told that wolf stories were expected by newspapers every so often and were always believed if the locale was sufficiently far away.'

How can we account for these differences in attitude? The pioneers' fear may have resulted from a hatred of the wolves that killed their livestock: seeing wolves kill domestic animals three times the weight of a man may have transformed their hatred into a fear for their own lives. The only losses by explorers were occasional raids by wolves on their food caches. And these men had greater contact with native people, who had little in their culture to suggest a fear for their lives from wolves.

'Think I'll move down a ways along the edge. The light's better now with that bit of moon. Maybe I can see if they've got a kill out there.' Lawrence moved off quietly. I watched his dim outline for a minute, then it blended into the trees. After a short time, I heard him howl. Before he had completed it, the whole pack was in chorus, still in the same place, strung out in the beaver meadow. If only we had had more rods in our eyes we could have seen what was going on. No single wolves howled back. They may have returned to the pack, or moved away out of hearing.

To some Indians, wolves had a special meaning. Among many

nations, including the Iroquois in the east, and the Tlinkit and Tsimshian on the west coast, they provided a 'totem' or figurehead for a whole clan. The Walla-Walla Indians of the Columbia River attributed supernatural powers to the 'great medicine wolf', who, in their folklore, was able to change people into rocks. So fundamental in the religion of some cultures was the wolf that the early Jesuit missionaries spoke of the devil, their adversary for the Indians' souls, as *ce loup infernal.*

All Indians did not grant wolves complete sanctity. The plains Indians draped wolf skins over themselves as they crawled within arrow-shot of the buffalo. Eskimos used wolf fur for trimming parka hoods, since it does not become frosty and stiff from their breath. But neither killed wolves out of fear. Samuel Hearne, describing wolf dens in the Arctic in the mid-1700s, said, '. . . yet I have frequently seen Indians go to their [wolves] dens, and take out the young ones and play with them. I never knew a Northern Indian hurt one of them: on the contrary, they always put them carefully in the den again.'

Lawrence slipped into the grass beside me. 'Seen nothing. Can hear 'em in the grass. Too bad the moon ain't full.'

We left the beaver meadow soon after that, even though we knew that if we stayed we might see the wolves at daybreak. But our job was to keep searching; being entertained by one pack wasn't helping put Xs on the map at the Wildlife Station. No wolves snarled at us as we stood up to leave, or looked at us with 'protruding eyes and lolling tongues'. No newspaper dispatch came out of this wolf-man encounter. There may have been one wolf on the trail ahead of us as we moved off; a branch broke and we heard a thud that sounded like a wolf bounding over a log. But he didn't tree us, not because the circumstances were any different from poor Thomas Conant's, but because we interpreted events differently — not as an attack, but as a peaceful meeting of two master predators, each living very different lives, each capable of co-existing with the other in our wild areas, if man chooses to respect the rights of other species.

6 Fast Water

'Go!' shouted the river, and with a mighty push on the stern of the canoe, the surging black water thrust us into the head of the rapids.

'Bow rudder,' shouted Lawrence, 'or we'll go down her sideways.'

There are few fast rivers in Algonquin Park. The Petawawa is one of them; born in the rocky highlands where melting snow and summer rains run down granite fissures and trace out the wrinkles in the thin mantle of soil, it gathers in first the Tim, later the Nipissing and Crow, glides into the Ottawa River where its waters sweep past the Parliament Buildings to join the broad St Lawrence at Montreal, and flow sedately under the last bridge at Quebec City and into the sea.

'Rock on the right,' I shouted to Lawrence. Instantly the canoe shifted in the white water, and the danger was past.

You can drink the water of the Petawawa. In the north it tastes of solitude and rocky shorelines and pine needles, but by the time it laps at the cement bastions of Quebec City it has been humiliated with sewage effluent and industrial wastes from each town and city it has passed — its biological oxygen demand raised, dissolved oxygen lowered, temperature raised, pH lowered, bacterial count raised, clarity lowered, conductivity raised, aquatic life depleted. Is it water any more? Or is it a liquid whose complexity baffles the biochemists?

The canoe slowed. We were in slack water again. Thirty seconds — that's all the time we took, instead of sweating over the portage for half an hour or so.

The days before we reached the Petawawa were hot. We travelled slowly across the large lakes — Burnt Island, Big Trout, Portal — fishing only a little and retreating into the shade of the pines during the hot afternoons. Sometimes darkness closed in and the first dew was forming before we got up from our evening smoke to begin our night's work. Nothing hurried during those days. Even a deer we came upon at the water's edge only lifted its head and stared at us. No wild fear, no snort, no stamping hoof, no jump to the trees with tail up. It was too hot.

We made Portal Lake by the fourth afternoon and lazed our way to its north end. Lawrence let out his trolling line, but the trout were languid, too. He only succeeded in snagging bottom as we came near the beach. Even in the shade of the maples, where we lay waiting for darkness, the heat seemed to rivet us to the ground.

In the openings among the maples, a few white pine saplings were beginning their long struggle upwards to reconstruct the forest of old. Grey blotches along their slender trunks revealed, however, that they would never wave above the maples. White pine blister rust, whose villainous spores sneak in on the wind, had already begun its work. Years ago their accomplice, man, had doomed these trees as directly as with an axe. Some unnamed timber entrepreneur, or perhaps a nostalgic pioneer, had introduced European white pine into North America, complete with its fungus disease, white pine blister rust. An ecological mistake! The disease swept through vast stands of native white pine, wiping out young trees and limiting pine regeneration. The crafty killer alternates between wild currant or gooseberry bushes in the early stages of spore development, and pine trees when it is ready to kill. Man, resenting being relegated to a supporting role, sets out to destroy the murderer. Crews armed with a herbicide and spray guns crisscross the more valuable pine stands, defoliating every currant bush they find. Now the herbicide (2, 4, 5-T) is cited as a possible agent of genetic mutations in animals and man. Arguments are raised in newspapers and courts. Partial bans are imposed here, total bans there. Have we just made another ecological mistake?

Portal Lake funnels into a short rapids which resonated through the woods around us while we waited for darkness. Magically, it began to stir our listless mood. Was it promising adventure? A break from routine? For the first time in days, Lawrence went off exploring. In a few minutes he came back for his fishing lures. For supper, rather than making only a minimal effort — cold beans and tepid lake water — we built a campfire and cooked lake trout. By dusk we were anxious to be off. A few miles of paddling and we reached fast water. Our whole pace changed.

It is impossible to travel slowly beside fast water: you feel impelled to rush, to keep up with the dark eddies slipping along the shore. They almost make you run. Pushed by the noise, pulled by the current, urged on by a feeling of uneasiness at the loss of hearing, you out-paddle the specks of foam on the water, and end the portages faster than you began them.

Lawrence tersely summed up the first rapids: 'Too steep.'

No conversation was required: our portage procedure was well established. On short portages I crossed first with my pack, returning part-way to meet Lawrence and relieve him of either his pack or the canoe. On long portages we stayed together, exchanging the canoe by merely lowering the bow to the ground while the new man came in from behind to take the weight. Sometimes the bow would slip forward, and the new carrier (not yet under the balance point) would stumble after it, trying to maintain his hold. When this happened, the canoe never followed the trail but took off into the bush, increasing speed until it struck a rock or stump and left you in a heap. It was amusing to watch Lawrence disappear this way.

We slipped back into the current at the foot of the rapids, keeping near the shore to reduce our speed. Lawrence steered and I scanned the darkness ahead with my light, paddle poised as a bow rudder. The current slackened and Lawrence began paddling again. We were beating the black eddies which glided along beside us. I drove my paddle into one; it gurgled and moved a few feet, then continued its business.

Before we had a chance to howl, the sound of the rapids behind fused with another ahead. We hadn't counted on all that noise; even if wolves were close, we had little chance of hearing them. But why should wolves bother howling in places where they are not likely to be heard by other wolves? Rapids may well create a vacuum in the forest, where wildlife seldom goes. How could songbirds stake out territories in such locations? Their songs would be obliterated. A male ruffed grouse would have little chance of 'drumming up' a hen. Under the best of conditions, ruffed grouse drumming carries little over a quarter of a mile. Near a rapids, the bird couldn't hear himself drum. Why would a deer go near a rapids and lose hearing, one of its primary means of defence?

Not all wildlife shuns rapids. Otters love them, sliding through the white water. But although the animals seem to be enjoying themselves, rapids have the same effect on otters as they do on man — they make him travel fast. The water ouzel, a songbird of the western mountain streams, has overcome the noise problem by its particularly loud song. Above the uproar its musical notes ring out, perhaps the loudest song among small birds. Despite the continuous roar, ouzels use song to parcel out the rushing streams into territories.

At the head of the next rapids we made a frightening error in judging the increasing strength of the river. Our lights were scanning the shore, searching for an opening in the trees, when suddenly we found ourselves being swept into the main current. Lawrence grabbed for an overhanging spruce limb; it slipped from his grasp. I made one frenzied stroke with my paddle, but the current almost pulled it from my grip. We had no choice; we were going down. White water rushed by the gunwales. Spray soaked our faces. A wave shipped over the bow deck and landed on my knees.

It was all over in a few seconds, and our canoe was in a backwater eddy at the foot of the rapids. We pulled over to the shore, and Lawrence drew a cigar from his pocket. 'Gotta have more respect for this river. She's got more power than the Tim.' The match flickered in front of his face. 'Shot a lot of rapids in my day and I never once swamped, but that one was plain luck. No rocks, no turns, just a fast chute of water.'

We sat in the darkness listening to the roar of the water. Wavelets rocked the bow of the canoe against a log. The waves, too, had spent themselves in that wild white race. They would have to gather new energy for their next dash a few miles downstream.

As recently as a hundred years ago the rapids of the Tim propelled the giant pines out of the park, the big squared timbers writhing in the white water. The shantymen camped below each rapids, guarding against log jams, jabbing at the timbers with their long pike-poles, jumping nimbly along the shore, riding logs through the slack water. They had missed the log that our canoe was bumping against, a water-soaked relic of the past. Logging was a hard and dangerous life for the men who took to the shanties each fall. Though it may be a romantic memory for those who look back at the early days of lumbering, the absence of the biggest pines in the east remains a constant reminder to the people who enjoy the park today of robbery in the past. Were masts for the Royal Navy important enough for the loggers to have appropriated them all, thus depriving the next generations of ever marvelling at majestic stands of pines six feet in diameter? Those trees took hundreds of years to grow, yet even today we persist in calling all trees 'renewable resources'. If man ever allows large stands of giant eastern white pines to grow to maturity, the first to have the chance to see them will be our great-great-grandchildren.

The canoe continued thumping against the log.

When the pines were gone, the loggers turned to spruce. Forest economics was rewritten in the early 1860s with the discovery that paper could be manufactured from wood pulp rather than fibres of cloth. The miles of lowland spruce suddenly took on value. J. R. Booth, lumber king of the Ottawa, saw it first, and recut his limits. But the haul roads were pushing into the wilderness by then, and as time passed, more and more logs jostled out on wagons over the corduroy roads. The days of the river drives became history, and the river began freighting a new cargo — people taking pleasure from canoeing fast water.

We carried our gear around another rapids a mile or so down river, this time landing and searching for the portage well in advance. Below the rapids, the river spewed us out into Catfish

Lake and freed us, for a while, from its grip. A cold display of northern lights brightened the night, throwing yellow-green beacons in long parallel lines across the sky. We howled for the first time since setting off, but no wolves answered.

The silence of the lake was intensified by the contrast with the roaring water. It was a silence of midsummer, without the spring peepers or wood frogs, with fewer bitterns or whip-poor-wills. The loons still called, though, making the lakes ring with their wild laughter far past the breeding season. Let us hope they are not singing a 'swan song'. Because of the loons' unfortunate physiological characteristic of storing large amounts of body fat, chemicals are accumulating in their bodies. During periods of high energy demand, such as the breeding season, these manmade molecules circulate through their metabolic pathways, potentially reducing eggshell thickness and reproductive success. Where do loons get the chemicals? From fish, crabs, frogs, and aquatic insects that inhabit polluted water. A hundred years ago, nobody believed that man would ever have the ability to wipe out the thousands of loons on thousands of wilderness lakes, many of which he may never even visit. But we can do it.

At the end of Catfish Lake, where a narrow chute of water cascades into a small lake, we almost made our second mistake. Lawrence got out to inspect the rapids and passed his verdict: 'Okay. She's fast but I can't see no rocks.' He climbed back into the canoe and kicked off. We picked up speed, but suddenly lurched towards the bank. Off balance, I reached out and grasped a cold piece of steel which was scraping along the side of the canoe. Lawrence quickly swung the stern to shore and held fast to a limb.

When I shone my light into the water, I saw a submerged square timber out of which stuck the long spike I was grasping. There was another and another, projecting just above the torrent. Part of a broken-down logging dam, the old timber with its spikes could have ripped the canoe from end to end.

It was becoming clear to us that the Petawawa was not ideal for nocturnal canoeing. After hauling the canoe around the old dam, we found a sandy point and boiled a billy of tea. The canvas of the canoe had been ripped where the spike hit, but the cedar planking was not broken. After a council meeting, we decided to go the rest of the way in daylight.

It was always a novelty to stick my head out of my sleeping bag in the morning and see a landscape that was completely new. I often held back the little surprise and tried to draw a mental picture from the shapes and shadows I had glimpsed in the darkness. This morning, I could tell by the warmth of the sun that I was in a clearing. I opened my eyes to see the sun streaming through a jack pine bough above me and I lingered in these pleasant surroundings, reluctant to crawl out.

Normally we think of the sun's energy in terms of heat, without which life would be impossible. But just as important, the sun is the source of energy for every movement of every animal, the fount of biological activity. We are only beginning to understand the pathways of its energy through living systems; this is one reason why some complete natural ecosystems — wilderness areas — must be preserved.

Wolves are at the top of one energy chain, consisting of a series of links which, if broken anywhere, would doom them to extinction. The chain begins with sunlight falling on a maple leaf, or a yellow birch or dogwood leaf. Most of the light energy is absorbed as heat, and some is reradiated back into the atmosphere. But about one per cent penetrates the leaf's epidermis to lodge in an internal cell. Instantly it is attracted to a chloroplast, one of the thousands of tiny green bodies that give leaves their green colouring. Introduced in the leaves to carbon dioxide and water, it reorganizes these to form a sugar molecule which, if not required immediately by the tree, changes to starch. The light energy has now transformed itself into a unit of potential chemical energy. This is whisked away by the plant's transport system, perhaps to a tiny axillary bud at the base of the leaf, and here it will remain even after the leaf that caught the sun's rays falls to the ground. Then, on a cold December day, a wandering deer may eat the bud. From here the pathways of energy branch in several directions, depending upon the needs of the deer. The chemical energy may be reconverted to sugar and burned for heat to keep frostbite from the deer's ear, or it could become part of a protein molecule, or be shunted to the liver and changed to glycogen. But this deer is well fed, and the energy is absorbed as a fat globule and stored beneath the hide of its back for a future need.

A month later, the deer wanders too near a family of wolves whose appetites are keen after days of unsuccessful hunting. Canine teeth rip into its back, and suddenly the fat globule is inside a wolf's stomach, attacked on all sides by digestive enzymes. The wolf may metabolize the fat globule, reconverting the energy in it to fat, or it may use the energy directly to move its leg, twitch its ear, or raise its head to howl. Then the energy is gone.

The deer has one other important energy pathway open to it, if it hopes to leave any clue for a wolf biologist to discover the state of its health just before it died. No doubt few deer care much for adding to the pool of scientific data in this way, but nevertheless, they may do so for metabolic reasons. Fat can also be stored in bone marrow, turning the marrow from red to yellow-white. If the wolves leave a deer's leg bone lying around, it will tell the story. A deer in good condition just before being killed will have yellow-white marrow; a deer on the verge of starvation, or perhaps dead and only scavenged by wolves, will have red marrow, having used up all the fat in its bones.

Lawrence was leaning over the upturned canoe at the water's edge. He was working pine gum into the ripped canvas and smoothing it with his pocket knife. That was the first damage we had done to the canoe, and somehow I felt we had betrayed a friend. Many tribes of northern Indians held the ancient belief that spirits existed in inanimate objects — sun, moon, rocks, and trees, even manmade artifacts. To ascribe such a spirit to a canoe is surely a more appreciative attitude than thinking of it merely as a collection of cedar planking, canvas, nails, varnish, and red paint. It seemed capable of reading our moods: when we stopped to howl, its staccato swaying would cease and the lapping of wavelets along its sides would fade into a whisper. Maybe it could hear the wolves, too. At another moment it knew to sweep boldly onto the crest of a wave. There it would hang, travelling at a terrific rate, while Lawrence kept it straight, and would finally plunge into a trough, as though enjoying the sense of fear that at any moment, if it slowed down, the swell behind would crash in upon it. In a playful mood, it often tried to trick me into a misstroke of the paddle by swinging up high on a swell, so that instead of biting into the wave, my paddle would skim the water and throw spray into Lawrence's face.

Our canoe was despondent in marshland. There the water lay still, as if covered by an oil slick. Wavelets that it loved to dance on vanished where the lily-pads began. The floating leaves slapped its sides and grabbed at its keel. If we stopped paddling, the canoe glided only a few feet and then gave up, held by the quagmire of vegetation.

Yes, the ancient belief of the Indians did have some truth; the canoe had a spirit — a projection of our frame of mind.

I rolled up my sleeping bag with an urge to feel the Petawawa's power in the daytime. We were on the edge of the sand flats that once formed the floor of a wide glacial spillway. That was ten thousand years ago. In places between the jack pines we could see burned stubs, evidence of the fire that is necessary to open jack pine cones. Because they cannot reseed themselves without the heat from fire, jack pines are living proof that forest fires have, for generations, been a common occurrence. We have been brought up to consider all forest fires bad. The imagery of Smokey the Bear has made us forget that some ecosystems only function if there are periodic fires. Ironically, not only Smokey the Bear but all his animal friends that surround him in posters and comic books — deer, grouse, raccoons — have a hard time finding food and shelter in mature forests that grow up when there are no fires.

For a long time, we believed that logging could take the place of fire in creating the young stands of vegetation needed for most species of wildlife. But the answer is not as simple as that. Clear-cut logging (cutting every tree in an area) may do so to some degree, by letting maple, cherry, and willow seedlings come back at the height deer and moose can browse. Raspberries and blueberries, growing in the sunny openings, fatten bears and fishers in the fall. Songbirds benefit from the edge effect, where forest and field meet. In fact, because plant diversity is greater there than elsewhere, the chances for many animals to find the types of food they need are also greater; therefore these edge areas tend to be particularly rich in many different species of wildlife.

But clear-cutting is not without ecological costs. The nutrients go out on the logging trucks, rather than being returned to the soil. Nurse logs, where seedlings find a rich rotting substrate, are removed. Often the tops and limbs of the felled trees are left to inhibit regrowth. And on any slope, the scars from heavy machinery start the processes of soil erosion.

Clear-cutting has mostly given way to selective cutting in the eastern mixed forests. Profit margins, export duties, and all the economic finery of the giant lumber companies does not allow time and money to fell substandard trees along with merchantable trees. While less disruptive to the forest than clear-cutting, selective cutting, too, detracts from park values. Often the forest canopy is opened only a little. Regrowth is sparse. No edge is created. Wildlife suffers.

Inevitably, any type of logging levies charges on the forest environment. On most forest lands, these are acceptable in view of our need for wood. But are they also acceptable in parks?

The little lake we camped beside pinches down into a fast two-mile rapids, where the white water churns over huge boulders and lurches against rock walls. Spray leaps into the air, gulping in mouthfuls of oxygen to meet the responsibilities of decomposition it will soon shoulder when it passes the mills downstream. The portage takes a short cut, and for a time the roar is subdued by jack pine needles and hazel brush. Then a short stretch of canoeing water, only sufficient to wet the bottom of the canoe, and another haul around a rapids. No chances to howl; it was best to get by the fast water in the daylight.

Ahead of us, farther downstream, the land dropped away. We pulled in at the portage sign and made our way down through the hardwoods around a forty-foot falls, with a shroud of spray at the bottom. Here was the scene of a tragedy during the river-drive days. As a prelude to our trip, Old Jack, the cook at the Wildlife Station, had recounted the fate of five men who drowned when their raft was swept over this falls.

To hear the full force of the river, I walked up along a log sluiceway long fallen into disrepair. A deep bass thundering made the old logs vibrate. The falls was singing no wilderness song; instead it was crashing out a warning: 'I am a reminder of the eternal forces of nature. I have thundered here longer than your written history, and may thunder here when man is only a memory, like the throngs of passenger pigeons that once roosted in the maples along my bank. While you can destroy me with dynamite and cement, you cannot stop the hydrological cycle of which I am a part. I will flow to the sea, and into the clouds, and over the soil, and through subterranean channels whether or not man jams me through turbines or robs me of oxygen or loads me with acids and phenols and wastes. But anger me and I will lash back. Strip the trees off my watershed and I will flood your cities. Load me with toxic chemicals and I will kill the fish you use for food. Dam me for electrical power and I will wait until you arrogantly build on my flood plains, and then wash you away. But leave me alone and I will add excitement to your wilderness adventure, at no cost, for generations to come.'

A last portage and we were at the wide beach of Cedar Lake. Here the river released its grip on us. We set out for the little railway village of Brent, a spot on the far shore. We hadn't learned much about wolves along the Petawawa, but we had learned that one facet of the wilderness experience is simply *adventure*, a basic need of man that many people find only rarely in our programmed cities. We had found it in fast water.

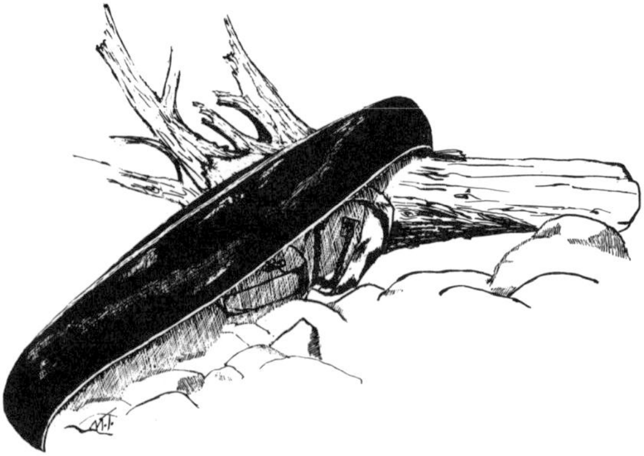

7
Big Crow

The nose of the Beaver aircraft dipped, and we banked down towards Big Crow Lake. The surface of the water spun upwards, and what had looked like tiny ripples changed to large swells. At the last second we levelled out, then — bump — we were down. Lawrence looked pale but relieved.

It was September. Along the shore, birches tinged with yellow stood out against dark curtains of spruce trees. September — and the final trip of our search.

Two weeks earlier we had paddled Big Crow, and watched a chorus line of northern lights flare its skirts to a pack of howling wolves. That was the twentieth pack found by the research team: Lawrence and I in the back country, Doug, Jack, and Ken in more accessible areas. For our finale we hoped to discover, by this second combing, if more than one pack hunted in the Big Crow country.

We had found twenty packs in the 1,100 square miles of our search — the same average density per square mile as previously recorded on Isle Royale in Lake Superior, and later in Superior National Forest. Most significantly, this density is the greatest on the breeding range ever recorded for wolves in North America. The following winter our results were verified by members of the team who made daily searches from the air. This winter census disclosed the total number of wolves in the packs, something we were unable to do in the summer: one hundred and eighteen, or a density of roughly one wolf for every ten square miles — the magic upper threshold that wolves only exceed briefly when following large bands of caribou in the north. Why, when food is plentiful, wolves do not increase beyond this number, is a secret we do not yet know.

Big Crow in autumn sunlight rivalled Big Crow cast in shadows by northern lights — blue water, white caps, and long foamy traces. Out of an extensive spruce bog, where the wolves had howled on our last trip, rose a hardwood ridge painted by frost in ochres and reds. Above the hill a hawk was climbing a windy circular staircase.

'Red-tail.' I heard Lawrence's voice behind me. 'See the glint through his tail feathers next time he turns.'

We beached our canoe on a sandy strip, and Lawrence set off to explore. I found a comfortable spruce stump to lean against and watched the waves crashing along the beach. As the sun left the lake, the wind was honed sharp by the evening coolness. A spotted sandpiper searching for aquatic insects teetered on a stick that rocked gently in the waves. In a few weeks the same bird would be balancing on a cypress root, hunting marine worms in the tidal flats of a Louisiana bayou. Suddenly it flew, propelling itself with quick erratic wing beats over the surface of the waves.

To be plunked down abruptly from the air on Big Crow hadn't seemed right. The feeling of remoteness takes time to be born. Part of the intangible pleasure of the wilderness only comes after hours of stiff paddling and tough portages. How much wilder this place had seemed when we reached it by our own labour! Had a canoeist been on this lonely lake when we landed, he would have had every right to be indignant at the raucous intrusion of the plane's engine. This had happened on other occasions, and we had taken the time to stifle the canoeists' resentment by explaining our mission. Without some air support and the occasional use of lumber roads, the accuracy of our thousand-square-mile census would have suffered, and the knowledge we hoped would assist in protecting the park's wildlife and help future visitors to understand more about their environment would have been incomplete.

Lawrence was returning; I could hear him sloshing through the swamp. A snowshoe hare suddenly bounded out in front of him, running straight towards me and thump — right into my leg. It rolled over, stunned for a moment, then quickly jumped up. Realizing I was another man, it zigzagged down the beach, clearing a deadfall, and finally dashed back into the trees. Its erratic style of running is a good example of adaptive behaviour. How did hares ever develop this trait?

When we recognize that zigzag running is part of the genetic makeup of hares and rabbits, evolutionary theory provides part of the answer. For this distinctive feature to evolve, hares had to be under pressure from predators. Rabbit is good eating for lots of carnivores: great horned owls, lynx, foxes. In the distant past, an aberration in genetic matter, a chance mutation, occurred to cause one animal to zigzag when it was chased. Perhaps a cosmic ray had struck one of its parents during a primeval period of solar bombardment, or maybe signals had become switched in the biochemical bonding of a chromosome. That such a specific trait could ever arise by chance is only made plausible by viewing it as one of countless genetic mutations that happened to the species over a time span beyond comprehension. Many mutations were lethal to the animal, and therefore were not perpetuated in the population. Others had no effect on the animal's ability to survive. But this one stayed on to help snowshoe hares survive today from the western forests of lodgepole pine to the taiga of the north. Under pressure from predation, an attribute like zigzag running protected hares against talons and claws, and assured a longer life span and more offspring. When the descendants also displayed the same behavioural trait, and survived to leave more offspring, the trait ultimately became indigenous to the whole population.

'Been wolves along the beach.' Lawrence sat down on a rock. 'Three of them, heading east. One's a big fella, too. Don't know if we're going to get any howlin' in, though. Getting pretty dark in the northwest.'

He pushed out in the canoe to get a better view of the sky. When he returned, he said, 'We'd better find some shelter; it's coming fast. There's a high stump washed ashore down the beach a ways.'

The rain beat us, and by the time one end of the canoe was propped up on the stump, there was little use getting under it. But we huddled in our cramped shelter anyway, leaning against our packs and listening to the rain beating on the hollow drum of our upturned canoe. Water splashed off the gunwales, and I tried to make myself narrower to avoid the steady drip. As it became dark, Lawrence's crouching form faded into shadow, and I could see only the glowing end of his cigar suspended a few feet away. Every now and then a flash of lightning would light up the face behind the cigar, staring at nothing in the blackness. Between flashes I found myself trying to recreate the face in my mind, then correcting the image when it appeared again.

'Water's running down my back and out the seat of my pants.' Lawrence broke the silence which had lasted almost an hour. 'Ain't going to get no howling in tonight. What do you say we head for the fire-tower man's cabin at the far end of the lake?'

We rolled out from under the canoe and switched on our lights. The swell was running high even near the shore, but the thought of a dry cabin was incentive enough for us to push out into the water. The lightning periodically set the hills and the shape of the lake in perspective and allowed us to gauge our progress. In the lee of a big rock bluff we swung across to the far shore. A light appeared ahead.

By the suffused glow from two coal-oil lamps I could see that the cabin was small. A few chairs, a table, a stove, and a pile of wood furnished the main room. A cot stood in a tiny alcove at one side.

'Ain't got no beds for guests, but you're welcome to the floor,' the tower man said as he seated himself beside the table. He was a tall, angular man, and the one uncombed shock of hair standing up on his otherwise bald head reminded me of a fire tower itself. Heavy galluses crossed his shoulders, more than ample support for a pair of baggy corduroy trousers. His arms hung at his sides, as if he didn't know what to do with them. He poured a cup of vinegar from a jug that stood open on the table. Taking a mouthful, he rolled it around for a few moments, then went to the door and spat it out.

'Howlin' at wolves, you say?' He drew a knife from his pocket and began carving a plug of tobacco. 'Lots around this summer. I seen one 'bout a month — no, two months back. I was comin' down from the tower and there it was standin' on the trail looking at me. I stopped and it jumped off into the bush and that's the last I saw of it.' He stuffed a big wedge of tobacco into his mouth. 'Hear them howlin', though — up the north end. They was goin' last night.'

'We seen their tracks,' Lawrence said, 'on the beach up there.'

'You boys land in the Beaver this afternoon? I thought Tommy was bringing me some supplies I ordered. Climbed all the way out of the tower and was startin' down the hill when I heard him takin' off. I come right on down then, anyways. That tower's a good climb. Hundred and three feet. Gettin' too old to run up and down her.'

'Used to have a tower myself,' said Lawrence, 'down near Bancroft. Not as far off in the bush as yours. Only a thirty-minute walk from town.'

'That so?' The conversation lagged. I pulled my chair nearer the open firebox door. Bullets of rain pelted against the metal roof, and the north wind hurled watery lances at the windowpane.

'See many people in here?' Lawrence asked.

'Few canoe trips go by, not many. This lake's kind of a dead-end street. Just swamps off to the east and north. Some comes in along the Crow River from Lavieille, but she's a poor river for canoeing. Lots of portages. Then there's a few make it down from Philip's Depot, but that's a straight walk — good many miles. I'm usually up in the tower there, anyways. Had one fella last week come climbin' up the tower. I was in it, but I hadn't seen this bunch, and all of a sudden I hear a banging on the floor hatch. First I thought the old tower was giving out — it's pretty old, you know. Well, then I realized it was somebody knockin', so I opened the door and there's a face starin' up at me, all red from the climb. He says, "Mind if I come in?" Well I figured if I said no, he'd just let go then and there, so in he came. Said his buddies would like to come up, too. Pretty soon there was eight of them up there, and me makes nine. Windows got all steamed up. Couldn't have seen a fire if it was goin' right beside the tower.'

Lawrence laughed.

'Put the kettle on there. We'll brew up some tea. What was you doin' up the other end of the lake all this time? Been dark a few hours now.'

'We was crammed up under the canoe. Thought maybe she'd only be a short one,' Lawrence answered.

'No sir. Saw the clouds rollin' up 'bout an hour after you landed. Whole northwest got dark. She'll rain best part of the night. Thought of radioing Pembroke and askin' for a couple of days off when I saw the storm comin', but I figured Tom wouldn't want to come flyin' back in here again the same evening. Anyways, I had a couple days off last month and not much to do. End up goin' fishin'. I get all the fishin' I want right here. No need to take days off to do the same thing.'

I took the boiling water off the stove.

'Some tea bags just over your head in the cupboard. Go all over the park, do you? Ever been down on White Trout Lake?'

'Couple times,' said Lawrence.

'Stopped in at the fire tower there?'

'Yep. Got a dandy spring — tapped right into his cabin. Always stop there for water.'

'Been talking to him almost every day all summer on the radio up in the tower. Never met him though. Tell me, is he a kinda short and stout fella? That's the way I got him pictured.'

'Well, not really,' said Lawrence. 'Fairly short, 'bout my height, but slim.'

'That so. Goes to show you can't judge a man by his voice.' The conversation broke off again.

'Tin cups over here on the table. Sugar, too; don't have no milk — I ain't been to the dairy today.'

I poured the steaming tea.

'Do much huntin'?'

'Used to do a bit,' said Lawrence, 'but not any more. Gave all my guns away.'

'I love huntin'. That's the bad thing about working in a park — you can't hunt. Moose huntin' is what I like best. Funny thing happened to my pal last year when we was on a hunt up north of Sault Ste Marie. We started out from our camp one morning —

been there two days with no luck, but there was a fresh snow on the ground, about a foot or more. We always split up and met again just before dark to head back. The day went just like the other two, not even a good set of tracks. At four o'clock I heard a shot over to my left. That was our signal to meet. My pal was waiting for me at the edge of a little swamp, and we struck out for camp. Old Willy was always away ahead of me — pretty near six feet six, that lad. Suddenly I heard him shout, "Hey, come over here, quick!" I started out on a trot, thinkin' he must have found some tracks or somethin'. When I caught up to him he was standing in a clearing and facing a muskrat. Well, the muskrat made a charge at Willy, snow flying. Will, he grabbed it and threw it back, but the rat came on again. Will was trying to aim his rifle before it got too close, but it was too quick for him. Finally he got his snowshoe on it and aimed his gun down and let go a shot. The rat and the end of the snowshoe took off. Old Willy was dancing around on one foot screamin' "I'm hit!" with a few other choice words sprinkled in. Wasn't till I got him sitting down that I could convince him he still had a foot.'

'That's a pretty good one,' Lawrence laughed.

The rain was only pecking at the roof now. It was well past midnight. I wondered if the story hadn't been embellished just a little. Whenever two woodsmen get together there are bound to be yarns, each man trying to outdo the other. The listener doesn't exist in the narrator's mind; he is no more than a sounding board, an excuse to practise the art of storytelling. If the story is presented as fiction rather than as a recital of facts, I suppose the narrator is free to use all the imagery and imagination he can, and truth must never be allowed to upstage drama. In the old days, a vast number of yarns were spun around campfires, or in little cabins like this. They became folklore, changing slightly from campfire to campfire, but kept alive by the bush tradition of a leisurely after-supper smoke. Now, as remote areas shrink, the folklore is disappearing. Most camps are connected to roads that lead to town, so evening hours of talk, once the sole amusement for woodsmen, have been replaced by more 'civilized' forms of entertainment.

'Never hunted moose much myself,' said Lawrence. 'Used to

hunt ducks plenty, though. Remember once near Fort Frances, I was hunting with an old crony of mine.'

I watched Lawrence's eyes as the image of the cabin and his two-man audience dissolved into another day and another event.

'We wanted to scratch a few bluebills, so we hiked in to Gillard Lake and slipped the canoe into a long bay. Oscar, my Labrador retriever, was with us. We found a sheltered spot and laid out a few decoys and then slipped under cover in the rushes. The ducks were flying, and we bagged four or five. Oscar brought every one of 'em back after long dirty retrieves. We were doing all right until a red canoe came putterin' around the point with a man and a big black dog in it. He paddled down the bay, taking his time and spoilin' a good half-hour's shootin'. Finally he settled himself in a little cove on the far side. Nothin' happened for about ten minutes, then two coot steamed into sight. My buddy says, "I like eatin' coot just as well as ducks." So we opened up and shot one and winged the other. Oscar was gone before I could make a grab for him. He didn't want to lose the bird we'd wounded, and even though he had worked hard earlier that morning, he was gaining on the coot. Well, the old coot was pumping away — neck forward like they always swim — heading right for cover where the red canoe was hid. They both disappeared, and in a few seconds we heard a terrible uproar. I was blowin' my whistle, but Oscar paid no attention. He was determined to get that coot. In a few minutes out steamed the coot and went on around the point. Then out swam Oscar around the point. The red canoe came next, and the fella paddled down to where we was sitting. "Do you own that dog?" he bellered as he neared shore. "Ya, it's mine," I sez. I could see he was pretty mad. "Well, look what he done to my dog!" His dog was stretched out in the bottom of the canoe, lookin' up with sad eyes. Behind his ear was a long gash. "All he done was growl a little as that mongrel dog of yours went by, and your mutt swerved in and tried to tear him to bits."

'Well, I quick offered to pay any vet bill before he became more violent. It was two hours before Oscar came back. Never saw that coot again.'

The storytelling lasted a good many hours. Then we rolled out our sleeping bags on the floor by the stove.

In the morning the sound of an axe close by wakened me, and soon afterwards the tower man appeared, carrying an armful of kindling.

'Good thing I shoved some logs under the cabin or they'd be pretty wet this morning,' he said. In a few minutes the fire was leaping and the tea kettle was on the boil. The door opened and Lawrence stepped in.

'Kind of wet out there.' He took off his cap and shook the water onto the floor. 'She's gonna rain awhile yet. Quite a wind, too. I was up at your tower and she's really blowin' on top of that hill.'

'Did you go up into it?'

'No. Don't much like heights any more.'

'Real view up there. Hundred and three feet she is. On good days I can see clean over into Quebec.'

After breakfast our host built up the fire again, and lit his pipe with the burning end of a splinter that had fallen part-way out of the firebox.

'Twenty packs you say? Couple of years back, rangers used to take fifty or sixty wolves out of the park each year. Thought they had to do it to keep them devils from cleaning out the deer. How do ya know there ain't going to be more and more of 'em now that they ain't trapped?'

At the time, this was a hard question to answer. We weren't certain then that the population would become stable. But now we can look back after the years of research and answer part of his question. During the seven years of the wolf study, wolves were protected and, as nearly as could be discovered, their numbers remained at a constant level. Prior to that, the effect of removing an estimated sixth or even a fifth of the population each year would probably have been negligible; the annual birthrate could easily replace this mortality. In the area of the park we studied, pups represented about half the population in summer and a third of the population in the fall and early winter, as indicated by howling responses and later observations. That's at least twice the number needed to offset the former annual kill by humans. That pups can represent as much as sixty per cent of a wolf population was demonstrated in a study of a heavily hunted but still expanding population in one part of Alaska in the early 1960s. The Alaska findings suggest that increased mortality in wolves may simply be offset by increased breeding, or better chances for survival of the young.

If wolves are able to maintain their numbers at a certain level, as the recent research suggests, what does this tell us about the effectiveness of the widespread control programs conducted in the past? The answer is that except in areas where intensive and continuous efforts were made to deplete the wolf population, the controls probably had little effect on the numbers. The bounty systems, which resulted in wolves being killed over a large area, are an example. As for the future prospects for control, the findings have serious implications for any management programs that are intended merely to reduce rather than to eliminate wolves. Since these animals can replace even reasonably heavy losses, if man really intends to control them he must risk exterminating them altogether.

The weather cleared in the early evening, and below the dark racing clouds an orange sun appeared low in the sky. With the promise of a good night for howling, Lawrence and I bade farewell to our host. We paddled through the narrows into Little Crow Lake and from there into the marshy Crow River. An osprey, perching on a tall stub near the water, watched us as we glided past. Summer was over for it, too; there are no fish to dive at when the ice forms. It must ride the thermal updrafts all the way to the Florida marshes, hoping to avoid contaminated fish and buckshot from those who still think hawks only rob chicken coops. Lawrence slapped his paddle on the water and the bird took off, slanting up into the wind as it cleared the tops of the stubs.

By the time we were far enough along to make turning back impractical, a wind-driven drizzle had set in again. When it was dark, we howled, but the noise of the wind and water completely drowned us out. At the foot of Proulx Lake we waited awhile under a big spruce, hoping for a break, but none came.

Two weeks earlier we had struck fear into campers on a point near the portage. In the afternoon, when paddling up the lake, we had noticed a woman and child setting up a tent. About three in the morning we returned, howling at first from the end of the lake. A few minutes later, a light appeared on the point. By the time we howled again, it was a roaring bonfire. Wolf legends all agree that when attacked by wolves, you should build a big fire; the cowards will slink around in the shadows but never come into the light. We discussed whether or not to stop and explain the howling that seemed to be coming from a remarkably strange place — the middle of a lake — but we considered the danger too great. If a canoe suddenly hove out of the darkness, we might have met a barrage of firebrands, the last desperate action taken by so many victims just before 'fangs and death'.

At the peak of the summer season thousands of people canoe in Algonquin Park, travelling among this great density of wolves with never a fatality. Not even a hospital case! Deer flies draw blood in Algonquin, not wolves.

While our census brought about no outcry that wolves were a threat to canoeists, it did result in other loudly voiced criticisms by anti-wolf campaigners. One of the criticisms was that the Algonquin country had become a huge breeding ground for wolves, stocking vast areas outside the park. This statement was frequently declared to be based not on conjecture, but on fact. However, examination of the number of wolves submitted for bounty from areas adjacent to the park before and after protection revealed no support for this argument. While emigration as a mechanism for stabilizing wolf populations still needs more investigation, a protected wolf population can remain constant without it, as a study of the island wolves on Isle Royale confirmed.

We hoisted our packs and our canoe for the last portage of the summer — up through the sodden hardwoods and down to Opeongo Lake, largest in the park.

'Pretty stiff offshore breeze blowin'. Hope she don't get too rough before we reach the far point.'

Lawrence's fears came true. As we rode farther from shore the waves increased in size. Black sky merged with black water. We were being propelled at a great rate through the darkness. Now and then we heard the drawn-out swish of a whitecap piling over itself. Next moment we were riding one, and for an instant white foam boiled along the sides of the canoe, disappearing suddenly into a trough.

'Put your light on up ahead. Should be nearing the point.'

I switched on the light. It showed a line of white foam. Ahead of us the waves were pounding against the rocks. The stern of the canoe shot around, helped by an incoming swell. A wave crashed over the side, then another. I jumped out to prevent the canoe from being swamped, fully expecting to plunge in up to my neck. But the water was scarcely knee-deep. Lawrence climbed out too, and we hauled the heavy canoe up on high ground.

'Boy, that was close. Been pretty bad if we'd knocked the bottom out of her on the last night.'

We pitched our tent, tying it between two birch trees for extra support, and unrolled our wet sleeping bags. The wind and rain kept up all night, and I lay awake, cold and wet.

This was the best way for our search to end. Had the night been calm, with a loon's cry echoing across the water and stars shining overhead, we would forever have pictured that summer as gently fading away. When portages are interminable and lakes rough — these are the times we remember more vividly; they contrast with the peaceful scenes and enrich our memories with a wild seasoning. I was secretly glad it was raining and that my sleeping bag was wet.

At dawn, Lawrence climbed out of our shelter to look for a dry pine stub. The wind still whipped the point, and a thick fog raced over the water. He kindled a fire in a depression in the rocks and we stood close to it waiting for the water to boil.

'Think we could navigate down to the end of the lake in this fog?' I asked.

'Well, might be hard, but I'd rather try than spend the day and maybe tomorrow on this point. If we go between the islands, we can use them as a guide till we pick up the far shore again.'

We drank some tea and set out. For a while the tops of the pines on the Graham Islands kept us on course. I placed the map and compass at my knees, partially under the bow deck to protect them from spray. Just after daylight the wind had swung around, forcing us to make a series of tacks to avoid running parallel to the waves.

After the last Graham Island had disappeared behind us, we entered a grey void, travelling directly into the wind. With no visible means of measuring our progress, I imagined we were moving backwards. Now and then the nose of the canoe would dive at the wrong moment and a wave would break over the bow deck and splash around my knees.

Land! The dark shapes of trees loomed out, and we altered our course towards them.

'Better pull in and empty the canoe,' Lawrence called out. 'Shipped a lot of water back here.'

We paused only long enough to return the water to the lake. Then, for a short distance, we kept close to shore, but a deep bay forced us to leave its protection, and again we were in the deepest fog. In a few minutes an island appeared on our left, and we moved into its shelter. From it we were able to pick out the shoreline of Sproule Bay. As we entered it the fog suddenly lifted off the water — at first only a few feet, then it disappeared completely. The tiny dock and the road out were visible on the shore ahead.

I thought back over our search, and what the howls of the wolves had told us. The truths we sought were something like the black eddies on the Petawawa — just when you thought you could hold them still with your paddle and examine them, they shifted away. We had added to the store of natural history about wolves in discovering that they frequent half-dry beaver meadows, and stay in confined home ranges all summer. We had made our contribution to the evidence that wolves do better at controlling their numbers than man does at controlling his. But the black eddy that slipped away was the answer to the question — how? Are pups not adequately cared for? Are they ostracized from social units? Are some adults forced out of packs? Only by watching wolf behaviour could we hope to find the answers. And the curiosity that came out of that whole summer's work close to wolves, without ever seeing one in the wild, was enough to sustain years of further study.

I took a last stroke, and rested my paddle across the gunwales. We glided to the dock. Lawrence and I shook hands. Our search together had ended.

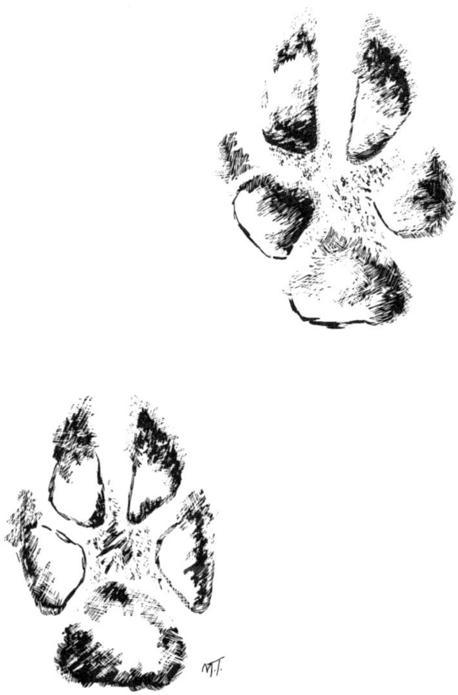

Wolves

8 Wolves

A grey-brown flash streaking across the headlights of my truck, and I had seen my first wild wolf after two summers of study. A flash in the opposite direction, and there went my second wolf. 'Studying wolves?' I'd occasionally be asked. 'Bet you've had some great eye-to-eye adventures!' Quickly I would explain that we work at night, but right away I could sense the disappointment. Scientific reasons aside, how I wanted to actually watch wolves!

But for Mary it was different. After one month of marriage and her first canoe trip, she had a great story to tell about what still stands out for both of us as our most picturesque wolf experience.

It happened at Fork Lake in Algonquin Park on a mist-shrouded morning when not a pine needle stirred. We were still half asleep when the wolves howled first — long-drawn-out musical notes rising and falling, their subtle harmonics borne softly on the still air. The wolves were some distance back from the far shore, and as Mary had never heard the wild woodwind symphony before, we slipped the canoe into the water and paddled very slowly towards the unseen orchestra. A low rocky bluff about fifteen feet high stood out through the mist and, resting our paddles, we glided to its base. Half an hour passed, a long enough intermission for the musicians, so I flicked the house lights with a howl. There was no answer. But the mist was blanching as the sun's rays washed it from above. Straight up was blue sky only partially veiled by a thin gauze of moisture-laden air. We lingered to watch the curtains rise on the day.

A sudden movement. Three wolves looked down on us from the rock. One jump and we could have been five in the canoe. They stood alert, shocked at finding people, not wolves. In a patch of full sunlight with wisps of mist stirring behind, they appeared phantom-like, moving yet not moving. Gaunt grey-brown animals, reddish behind their ears, thin-muzzled, and with long graceful legs. Fierce? Not any more than large curious dogs. Their expressions seemed almost pleading. But why should they plead with man?

For about thirty seconds all five of us remained motionless. Then the largest wolf whirled abruptly, almost knocking into the other two, and they vanished. The mist rewrapped itself around the rock, and we were left with the sensation that the wolves had never been there.

Every brief encounter with wolves is like an ounce of gold in the trust fund of knowledge about their ways. The appearance of the three had been a demonstration of the species' directional precision in discovering the exact location of sound. But to strike gold we needed more than thirty-second glimpses. A combination of our howling technique plus a large share of good luck did provide longer-lasting opportunities to watch wolves. Helping, too, was a degree of predictability in the movements of packs. After leaving their natal den in midsummer, wolves take up temporary residence in one of the several established rendezvous or resting sites within their home range. They may stay here while the pond lilies dapple the dark water of the swamps. By the time cicadas drone all day and bullfrogs rumble all night, the pack may have moved and the pups be playing perhaps three or four miles away. When the blueberries are ripe and a few maples on the edges of the clearings are prematurely red, the pack may have returned to the first site, or settled in yet another. After that, by the time the first frosts brown the bracken fern in the exposed forest openings, there is less chance of being able to predict their movements. By then the pups can travel as well as the adults, and packs lie up only briefly near the sites of their kills.

Two aspects of wolf behaviour are relevant to the question of how wolf populations are regulated. Both require daytime observations. Are wolf packs antagonistic towards each other? Does antagonism occur within packs either among adults or, perhaps more important, among pups? If among pups, we might discover the roots of social exclusion, and this might explain the disappearance of pups either through dispersal or through death.

On one of the occasions when wolves granted me permission to watch their undisturbed play, they accepted me perhaps as far as wild wolves can ever accept man. For sixteen days I sat on a ridge about 250 yards away from the wolves, and although most of the time they could see me, they went about their business undisturbed. Each morning I would arrive at daybreak, then leave just before noon when things quieted down, and return for an evening performance. At times different people accompanied me and they, too, were accepted as long as they played by the rules: walk slowly and sit quietly; no sneaking behind trees or loud talking — the behaviour appropriate for an old-fashioned prayer meeting. But these events were much livelier, since wolves appear to enjoy just plain fun.

The night I found the pack, a sharp wind was swaying the tree tops and turning over the leaves. I was about to give up howling, when a gust blew back a faint reply. I took a compass reading, then walked half a mile farther along the logging road and checked again. This gave me a perfect cross-bearing which placed the pack on the edge of a large clearing. The following evening, just when the sun was highlighting the woods with an intense yellow-orange light, I walked to the clearing and cautiously peeked over a rise. Ten years or so earlier, a forest of mixed red pine and maple had burned there. Plant succession had been slow, and when I first saw the old burn it was dotted with low blueberry bushes and bracken ferns, with a crisscrossing of burned trees and stumps. One side of the clearing rose to a hardwood hill; spindly black spruce surrounded the rest. In the middle was a small marsh, where earlier in the season redwings had nested and bitterns stalked.

I was not the only spectator; half a dozen bluebirds flashed among the few standing burned stubs, catching insects on the wing. They broke the rules about moving slowly, and their call notes carried most of the way across the clearing. The wolves tolerated this; they have never been shot at by bluebirds.

My field notebook records the action: 'On my first scan I spot an adult wolf about one hundred yards away, moving slowly, head down and ears up, along the edge of some leatherleaf. Now and then it stops to investigate the area around its feet. It is moving away from me — doesn't know I'm here. The sun is shining on the tawny backs of its ears. It comes onto the plateau of blueberry bushes, still moving very slowly. Stops and looks all around. Sits down. Stands up and snaps at a fly. Lies down, but still watching. Up again and trots into the dry marsh. Can't see it.

'Three minutes later — I see two adult wolves on the edge of the plateau. They move slowly with tails low. One suddenly jumps up onto the shoulders of the other. They run a few yards, still jumping, darting in and out in mock attack. One wolf rolls over onto his back and the other stands above, pawing him. Then he runs down into the marsh and the second wolf follows slowly. Both leave my field of view.'

A brief display but a significant one. With no sign of aggression one wolf had expressed its subordinate position by rolling over, the other had reasserted its dominance by standing above the first. When critical decisions affecting all members of the group must be made — decisions on such matters as when and where to hunt — a social hierarchy is essential. Had these animals resorted to fighting to establish who was boss, their behaviour would have run counter to the evolutionary goal of self-perpetuation.

Man, too, normally uses subtle means to establish and enforce his hierarchies although, unlike wolves, he sometimes resorts to intertribal warfare, such as civil wars and rebellions. We can observe these subtle group mannerisms at any cocktail party or office lunch room, or meeting of the board. Although adult members of our sophisticated society don't lie down on the floor, only high-ranking individuals dare grimace in disapproval when another member of the group passes a contrary opinion, and only certain people dare interrupt certain other people.

But the show wasn't yet over for the night. Five minutes later one of the two wolves was back, and lay down beside a stump. No sooner had he appeared than a number of wolves began howling from a short distance off to the northeast. Immediately he sat up, put his head back, and answered. The second wolf, just out of sight, joined in a duet which lasted about two minutes. Then they both trotted across the plateau in the direction of the howling, pausing occasionally to jump at each other in friendly by-play.

Still all good-natured. Where were the snarling slavering beasts of the legends?

My notebook continues: 'I see one wolf racing across the plateau. It comes to an abrupt stop, waving its tail violently over its back. It meets four more wolves, running to it from the direction of the howls off the burn — three partly grown pups and one adult. All five wolves form a knot of jumping, tail-flicking animals. They run in a group, still wrestling and rolling over each other. The one adult that came with the pups now stands to one side in an attitude of parental approval. They run towards the dry marsh — the grass is longer here and partially hides them. Occasionally a pup makes a spectacular leap straight up, exposing its whole body — sometimes two or three times in succession. They move into some tag alders and I can't see them any more. Getting quite dark.'

The wolf pack, four adults and three pups as I counted later, allowed me to watch five such 'greeting ceremonies'. The others took place when single adult wolves returned to the pack in the early mornings after a night's wanderings. All these reunions seemed, at least to human eyes, to radiate unrestrained pleasure. While serving to reinforce social bonds, these exuberant displays also provided dramatic evidence of the strong ties between the members of the wolf packs.

When it was too dark to see even shadowy movements, I left, walking slowly in case wolf eyes were watching. A mile away I rolled out my sleeping bag under a pine and, too excited to sleep much, waited for daylight. The pack howled during the night at 2:40 and again at 3:10. At 4:10 a single wolf howled quite close to my pine tree, driving away all possibility of sleep.

At daybreak I was at the theatre again for Act Two, accompanied by Doug Pimlott, director of the study, and his son, whom I had alerted earlier to my possible chance to observe wolves. When we looked over the ridge, one adult wolf was lying at the base of a stump, while the three pups played together a few feet away. The adult watched for a while, then lost interest and put his head on his paws. A wolf howled from the trees just to the east of us. The four wolves jumped to their feet and rushed out towards the sound along the edge of the plateau to join the homebound adult. All united in a greeting ceremony similar to the one the previous evening. When it was over, one pup stretched out in the grass, another sat down beside him, and the third disappeared into the dry marsh. Within minutes they were back together romping and playing like uninhibited children. Sometimes their play lasted for hours. My notebook describes one of these characteristic frolics:

'7:10 — One pup, tawnier behind the ears than his litter mates, is sitting at the edge of the clearing. A second pup appears from the trees behind and runs to him. Tawny stands up. The two pups trot a short distance. Something in the grass attracts them. Tawny stands still — the other pup is half crouched. The latter holds up its paw, then pounces. A mouse? They continue and suddenly Tawny jumps at his companion.

'7:15 — They begin to play, running among the stumps, jumping, rolling over. Tawny crouches behind some tall ferns. His companion cannot find him but wanders near and suddenly Tawny springs out and the two roll together. They jump up on a log, biting each other's necks. The second pup grasps Tawny's lower jaw and twists sideways. They jump down, chasing each other.

'7:30 — Both pups lie down about fifteen feet apart. They appear unconcerned with each other.'

Could anyone, after watching such joyful antics, deliberately kill a wolf for 'sport'? If wolves were less secretive and their family games well known, our sympathetic society would undoubtedly have included them sooner under the net of concern we extend to whooping cranes and baby seals. Merely the similarity of their behaviour to that of dogs would have strengthened our compassion; we might as callously put strychnine in Fido's dog bowl as lace it into the carcass of a deer. Indeed, the wolf is the ancestor of many breeds of dogs that we play with, love, and even invite into our homes. We did not breed play into dogs; ever since their ancestral beginnings it has been an important means of communication. When Lassie wags her tail and rolls over to let you rub her stomach, she is telling you in her age-old way that she knows you are the boss. At the same time she is avoiding any need for you to remind her more forcefully of your dominance. She is strengthening the social bond when she runs with you and brings back the stick you threw. Her play is merely an inborn reflection of the past.

For both wolves and dogs, the important period of socialization takes place when they are three to ten weeks of age. During this time play provides the mutual interaction necessary for them to form a bond with their own kind, a bond that later results in both co-operative hunting and care of their young. Wolf pups, through play, learn the individual characteristics of their litter-mates and other wolves in the pack; the groundwork is laid for the social hierarchies that enable wolves to resolve disputes without fighting, and perhaps even to solve the problem of regulating their own numbers. But despite my careful observations, I was unable to detect any dominant-subordinate relationships among the pups. The outward manifestations of these must therefore show up after ten or twelve weeks, the approximate age of the young wolves I watched.

Wolf pups also benefit from play through its contribution to their later hunting behaviour. The pup that crouched behind the ferns and jumped out at its litter-mate will some day use the same motor movements to spring at a deer. Similarly, hunting depends on pursuit and capture, common elements in play as pups chase and knock each other over. Hunting, however, is not entirely learned from play, and requires actual participation with adults. An incident involving the wolves at the Wildlife Research Station served to make this clear. These wolves, captive since puppy days, were fully accustomed to both humans and other wolves. As pups they had played together in semi-wild conditions, but had never seen a wolf make a kill. In early May, one of these wolves, named Kit, who was allowed some freedom within the research area, was trotting along the shore of Lake Sasajewun when he spied a deer. He gave chase. The deer, confused, bounded into the water and began to swim. Halfway across the lake it came to rotten ice which it could neither climb up on nor push its way through.

Kit, however, was able to pull himself out at another indentation in the ice, and ran to face the trapped deer. But when he got there, he was not sure what to do. He stood facing it, confused, and finally trotted back to shore. In endless hours of play with his four litter-mates, he had perfected the chase, but he had not experienced the final act, the kill.

The wolves on the burn, after a brief intermission, put on a dazzling Act Three, a once-in-a-lifetime performance. A wolf-bear encounter! It was too exciting to take notes; I wrote it up later: 'When I first saw the bear it was running directly towards the wolves, apparently having either seen or scented them. The wolves jumped up and to my amazement ran at full speed towards the bear. They met, wolves jumping around the bear and the bear lunging back at the wolves. They all ran together, the bear chasing one of the wolves while two other wolves chased the bear. In a hollow the bear halted, and the wolves circled it. More jumping and lunging. The bear seemed clumsy compared to the agile wolves, although one wolf narrowly missed being struck by a powerful swipe of the bear's paw. The entire cast began running again, reached the trees and disappeared. Moments later the three wolves walked back into the clearing and lay down. Through it all there was no actual contact between bear and wolves.'

Whether the wolves were being antagonistic or playful was impossible to tell. Their actions seemed to be those of play — chasing, jumping, darting in and out. Perhaps they were using the elements of play for a different purpose, to distract and lead off a potential threat to their pups.

Among other people who have watched wolves and bears exchange greetings, Adolph Murie in 1944 described a similar encounter between a wolf and a grizzly bear, which he had witnessed in Mount McKinley Park, Alaska. But wolf-bear skirmishes can sometimes be serious. Paul Joslin, another member of our research team, discovered a dead wolf at a den. A mother of pups, the wolf had eleven broken ribs. The hair around the den indicated that she had probably been killed by a black bear while defending her young, which were only a few weeks old.

During the time I was observing the wolves on the burn I made my temporary camp near an abandoned lumber shanty two miles away. After a few days, the pack moved over to the north edge of the clearing and in doing so discovered my presence in the grandstand seat on the ridge. For hours every day I sat clearly exposed and in full view of their rendezvous site, but they seemed to care little. The wolves watched me howl and I watched them howl. It was a mutual study, unique for them, too — a chance to observe a human who wasn't plotting ways to collect a bounty. As a gesture of appreciation I dragged a dead deer, which had been hit on the park's highway, out onto the burn. Contrary to legend, I was not set upon because the smell of deer's blood was on my arms; the wolves thanked me by allowing me to escape unharmed. To avoid disturbing them too much, I had picked a hot afternoon when they were all sleeping in the spruces.

The wolves weren't greedy. They waited a couple of nights before eating the deer. Perhaps the human scent around the carcass put them off. Once an adult wolf strolled casually up to the deer, jumped back, trotted quickly a few yards, turned sharply again, and then returned the way it had come. But eventually they cleaned up the deer.

The wolf-play dramas, in failing to reveal any hierarchy among the pups, indicated that, until late summer at least, social exclusion is probably not the cause of pup losses. Less conclusive was the evidence that might have answered the other important question: is there antagonism between packs? During my observations the pack on the burn was joined by another pack, but the wolves forgot about my poor night vision; the behavioural rites, if there were any, were performed in the dark. The circumstances were recorded in my notebook:

'20:00 — A pack responded to my howl from half a mile northwest of the burn. At 20:20 another pack howled from the burn itself. To discover if the same pack gave both howls, I advanced slowly towards the first howl. At 21:20, an hour later, I was almost at the original location of the first pack and, as nothing had happened, I howled. The pack responded from the trees just in front of me. Another pack answered from the edge of the burn. At least two wolves howled from positions in between.

'At daybreak both packs were still there, this time in the trees at different edges of the burn, with individual wolves again howling between. A few minutes later four adults and all three pups were in the open, but the other pack did not come out. Rain began, and one by one the wolves returned to the trees.'

That evening only one pack answered my howls; the other pack had left again. I had been close enough the night before to hear any voiced antagonism and was sure there had been none. But who were these wolves that had come and gone? Were they only absent members of the same pack? On previous occasions I had located another group, which seemed to stay at least three miles away. They were nomadic, probably without pups. But not knowing either the relationship of the two packs or the degree of separation over a sufficiently long period of time, I was left with only tantalizing theories.

Eventually the wolves must have accumulated enough observations and pulled out, leaving the bluebirds dipping between the stubs and me howling only to the wind. Doug and I searched the trees around the burn for evidence of kills, but found little. A calculation of the food requirements of the four adult wolves, made on the basis of various known values such as the amount of meat needed to sustain wolves, indicated that over the sixteen days they would have required about half an adult deer and five fawns, or a similar ratio approximating 330 pounds. The pups needed food as well. The deer we supplied contributed part of this, but where were the other remains? At my abandoned lumber camp a buck in its prime stamped its hoof and snorted at me almost every day. But with a density of ten to fifteen deer per square mile, the wolves had lots of choice.

I have returned to the burn a number of times since, but wolves no longer play there. Aspen trees have taken over, and from my lookout I can see only a few feet through the green foliage. Such are the dynamic processes of the forest. Soon the ridge will grow red pines and maples again, and nobody will suspect it once grew wolves.

It was on another burn thirty miles away and eight years later that Mary and I, still trying to pin down the same questions about wolf pups and social behaviour, obtained front-row seats for another wolf drama. With us was Jenny, our six-month-old baby, who was soon to tempt the wolves with helpless young human flesh and live to tell the story.

We were making another census of the wolf population in one section of the original study area. Because we had help from student assistants this time, we were able to take a few days off to observe a pack of wolves whose rendezvous site was in a particularly open area. Our cross-bearings placed a pack about half a mile out on the blackened flat terrain of an extensive clearing. At dawn, loaded down with binoculars, camera, tape recorder, and spotting scope, we approached slowly, stopping to scan with binoculars every few steps. A fire had swept through this part of the coniferous forest some fifty years before. The logs that had crisscrossed the land immediately after the flames now grew blueberries in the strips of their rich rotting humus. Speckling the burn singly or in clumps of twos or threes were large spruces and balsam firs that had profited from the nutrients released to the soil immediately after the fire. A red squirrel chattered from the top of one of these, a reliable sentry sure to alert the wolves. If that was its intent, it succeeded, for where we expected to see wolf pups rolling and playing we saw only an alder-choked depression through which meandered a tiny stream. The wolves, we surmised, were well concealed down in the alders.

We pitched our tent about a hundred yards from a widening in the alders and decided to wait the wolves out. Low clouds loitered in the upper branches of the scattered trees. Mosquitoes, elated at the windless flying weather, quickly closed in. Jenny, resorting to the only defence she knew, uttered howls which rivalled those of the wolves themselves. We did not expect any wolf activity until evening; nevertheless, we took turns watching the burn. Whoever was off duty wandered back along our route to gather blueberries and collect pine knots. These knots were all that remained of tree trunks which had returned their nutrients to their source. In our fireplace the next winter, the resin from the knots blazed yellow, giving up energy that had been caught by pine needles on sunny days long ago.

In most of the Algonquin country fifty years after a fire, aspen leaves would dapple any sunlight that reached the ground. In this half-light, maple saplings on the hillsides and spruce in the lowlands would struggle for dominance, their spindly arms striving to overtop each other and reach the sun. But not here; aspen was absent, and cherry bushes with the thick stems of old age had grown only to a height of a few feet. The reason plant succession had ceased was recorded most clearly by the scattered spruce and balsam. Almost all these trees were limbless under ten feet, with only browsed stubs for lower branches. Moose had altered nature's usual pattern when they discovered at this place the two prime requisites of a perfect wintering ground — dense coniferous shelter adjacent to the burn, and new growth at the edges and out on the burn itself. Deer wintered here, too, browsing not only the cherry shrubs but new growth on the cutover hardwood hills immediately to the north. In an interesting ecological feedback, moose and deer were maintaining the forest in optimum condition for themselves rather than merely responding passively to forest succession. Helping them may have been a fortuitous lack of aspens close to the burn, or perhaps other subtle features such as the sandy soil. Nevertheless, to hold the forest back so long was a remarkable feat of environmental manipulation.

It was dusk before the actors arrived on stage, with a backdrop of fading colour and merging shapes and to the violin accompaniment of long-horned grasshoppers rasping their wings together rhythmically. I howled once from the brink of the alder depression and immediately the clearing came alive with a reverberating reply. The wolves were fanned out in front of us, none of them further than ten canoe lengths away. Across the depression three pups streaked through the blueberry bushes, racing towards an adult down in the alders. Then occasional whining and yapping. Were they arguing over a bone? I carefully adjusted the microphone parabola to record the dog-like sounds and howled again. Immediately we heard a rustle of leaves and a swish of running feet. Three grey shapes bounded out of the alders towards us — we had not been seen, since we were crouched down operating the tape recorder. Two of the shapes came around one side of a spruce tree less than thirty feet away, the other swerved around to the right. I jumped to my feet. The first pair veered off, hardly breaking stride, but the third wolf, almost on top of us, fell backwards. As it scrambled up again, I switched on my flashlight. It faced us and stood perfectly still, like a little boy caught in the act. It took a step sideways, never lowering its eyes from the light. For a full two minutes its nerve held. Then it walked briskly, almost arrogantly, back into the alders.

We sat at our lookout until the early morning hours, recording the best tapes in our collection. Perfect conditions — a thin mist, not even a whisper of wind, and a willing pack very close. Because the mosquitoes also found perfect conditions, we were forced to suspend the microphone well away from us, to avoid recording any sounds of buzzing and slapping or the occasional whispered compliment about their ferocity. Throughout the night Jenny slept without comment in a pup tent with wolves less than a hundred feet away, sometimes closer to her than her parents. As testimony to our faith in wolves, it wasn't until some days later that we realized neither of us had even thought of the possibility of danger to the baby. Had it all happened in pioneer days, however, the three wolves that almost ran us down would have been attackers, and we, the terrified parents, would only have survived by sitting high up in a tree all night, clutching the baby, while the hungry wolves howled around its base. In truth, we had almost been included in a wolf greeting ceremony. With no malice, the 'attackers' only wished to reunite with their pack members before some of the group left for the night's hunt.

The tape-recorded howls we collected that night served us beyond the pleasure they provide when we play them to the flicker of firelight on our living-room walls. Determining wolf densities by howling has the unfortunate drawback of telling us little about possible changes in the size of packs. By analysing the sounds of recorded howls, however, I hoped to discover pack sizes as well as to find out whether pups were present in each pack. The shorter, more treble and immature howls of pups are sometimes overridden by adult yapping and howling. Sound analysis helped verify that all twelve packs we located that summer produced a litter of pups. Since the number of packs in the 600-square-mile area of our census was about the same as in former years, this high production of pups confirmed that population regulation in wolves is tied to the disappearance of the young. Had they all lived to occupy the same range, the next summer Algonquin Park would have been knee-deep in wolves. But it wasn't.

At daybreak, after a few hours' rest and with Jenny still sleeping soundly, we returned to the edge of the alders. The wolves answered my first howl — all in the same place. Then two pups appeared, running through the blueberry bushes across the depression. One saw us and came to an abrupt stop. It looked us over for at least three minutes before running on. For the next two hours both adults and pups wandered out into the open, sometimes running together, sometimes lying down. At first they were visibly disturbed by our presence and stood watching us for long periods. After about half an hour we gained their trust and they ignored us, but they engaged in very little actual play and kept secret any social relationships. Once five pups, all a similar light grey-brown, were out in the open together, and although we saw only three adults, later sound analysis showed that the pack consisted of five adults. By half-past eight the mist had burned off, and the pack retired to the alders.

That evening the curtain rose about the same time, and the performance was remarkably similar. The entrance came after nine simulated howls. A stick broke at the edge of the alders, then whining of a pup, and an adult wolf raced towards us. My light was ready and the wolf stopped twenty feet away, transfixed as before. The tripod and camera were all set up for a time exposure, but in handing the flashlight to Mary I dipped the beam off the wolf's face. The wolf was gone immediately, this time running around behind us rather than straight back to the alders. We didn't see the pup; it, too, must have returned to the trees.

We howled over and over again that night, but the pack had gone. The adult and pup must have been stragglers which, after their encounter with us, moved up the alder depression to travel with the others. We howled for them the next two nights, but they were completely out of hearing.

The wolves did leave some valuable information behind: their droppings. Our search on a hot afternoon a week later was thoroughly gratifying — sixty-six scats, as they are called. While the science of scatology, a legitimate interest of biologists, is sometimes scoffed at, the information contained in scats can often be gained in no other way. Through microscopic examination of the scale patterns on the hair found in these wolf scats, we determined that close to two-thirds of the animals' diet while they were on the burn was beaver, with deer and moose making up the balance in approximately equal proportions. Some scats contained winter hair of deer, which differs in scale pattern from summer hair, indicating that the area had been used at least periodically for some time.

This high percentage of beaver and low percentage of deer contrasted with the food habits of wolves in the region ten years before, when studies showed that deer made up eighty per cent of the wolves' summer diet. It reflected the great decline in deer that had taken place all across the northeast sector of the species' range during the late 1960s. This decline has occurred not only in regions inhabited by wolves but also in places such as the Atlantic Provinces and New England states, where wolves are not to be found. Whether beaver and moose will sustain the wolf population in Algonquin Park as lavishly as deer have in the past can only be determined over a period of time. Investigations on Isle Royale have demonstrated that moose and beaver can sustain wolves if moose are sufficient in number to provide almost all the food wolves need in winter when beaver are inactive under the ice. The ups and downs of wolves and their prey are part of an age-old pattern; no prey species can become extinct without either spelling doom for the predator or forcing the predator to find other prey species. Natural selection, driven by the predators themselves, normally keeps the prey species one jump ahead.

We returned to this burn the following year, and the wolves were there again. As we collect more tapes, we begin to recognize features in the howls of different members of the pack and may soon be able to follow the fate of individuals in later years. But that will be possible only if moose continue their good husbandry and keep the forest from encroaching.

With renewed evidence that the key to the regulation of wolf populations is in some way related to the fate of the pups, the next summer we observed wolves in an entirely different environment. Kluane National Park lies across the high St Elias and Kluane Mountains in the southwest corner of the Yukon Territory. A few years earlier Mary and I had been involved in efforts to preserve this land as a national park and had hiked many of its valleys, describing and photographing its wildlife and biotic features. Although we were not specifically looking for wolves, our eyes always searched for them along ridge tops and river flats, but despite Kluane's open tundras we had not seen any. Nor had we during three previous summers in interior Alaska where, in the course of other studies, we always watched and listened for them. Once we saw their tracks behind our Alaska cabin. They had come up our valley, following the gravel bars of the creek, but passed silently, and when we discovered their tracks they were probably miles away. The only wolf we had ever seen in the northwest was in the autopsy room of the Alaska Department of Fish and Game in Fairbanks, an animal that had just been brought in for a bounty. Its frozen eyes were as uninformative as the stainless steel table on which it lay.

So we were greatly excited when we learned from Tom Cottrell, a graduate student, of a den he had found only a day's hike off the Alaska Highway in the newly established park. It was not a tundra den as I had hoped, but rather in a valley, dug into the bank of what had been a braided stream. Although spruces surrounded it, out in front of the den the old gravel wash was colonized only by a low-growing mat of yellow dryas and scattered small spruce trees.

The solitude and beauty of the wilderness setting redeemed the lowland site and lack of exposure of the den. An amphitheatre of mountains surrounded the valley which, on that first calm evening, bounced the musical howls of the wolves back and forth, creating a resonance new to our ears. The spruce forests ended two thousand feet up the flanks of the mountains, giving way to a light-green band of low subalpine willows and dwarf birch. Strapped above this was a narrow belt of tundra proper, then jagged rocks, snow patches, and talus slopes. Mountain goats, tiny animated specks, were pasted onto the high rock cliffs far beyond the reach of canine teeth.

When we first approached the den, we were obstructed by a grizzly calmly grazing on vetches along the old mining trail. It was less than a hundred yards away when we rounded a corner and saw its big brown rump ahead, but the wind was towards us, and the bear wasn't immediately aware of its only enemies. With nothing more than stunted balsam poplar trees within sight, and with Jenny riding in her carrier on Mary's back, we conceded it the trail. At first it was moving the same way as we were, and as it was only about a quarter of a mile from the wolf den, we thought we had just bought tickets for a wolf-grizzly encounter. But, still unaware of us, the bear turned and began grazing its way towards us, ambling faster than before. We retreated to a stand of climbable spruce trees, but when the bear reached the point of our farthest advance, it simply vanished into the poplars. Grizzlies fear man. It is only the bizarre and unusual event of an attack that makes headlines. Thousands of peaceful encounters go unreported, but still the very name grizzly strikes fear into many people. Not that you shouldn't be cautious in grizzly country. Experienced travellers rattle a 'bear can' — stones in a billy pail — as they walk through the bush, for it may be dangerous to confront a grizzly by surprise. But grizzlies wish to live peaceful lives, searching for berries and digging out ground squirrels. They have no special taste for human flesh.

After the bear had yielded the right of way to us, we advanced to the spruces just in front of the den. A chocolate-brown wolf pup, darker than those we had watched in Algonquin, was standing at the edge of the trail. I howled and it threw back its head and answered — short and yappy. A second pup howled with it.

A minute later an adult wolf stepped out and, not seeing us crouched down in the spruces, trotted in our direction. He was rangy and looked big-headed because of his slender body. From about 150 feet he howled, at first very bass, then treble. Both howls were short with a low volume; their carrying distance, even in the calm of evening, could not have exceeded a few hundred yards. They were undoubtedly meant for the wolf he expected to find. He came on slowly, nose up, receiving the first scent signals that we were not wolves. Then he saw us and stood motionless while I snapped one underexposed picture. At the click of the shutter he turned and loped into the trees.

The site of this den was different from all the valleys we had explored; small mammals, super-abundant elsewhere, were scarce here. To the north, snowshoe hares, at one of their cyclical peaks, had gnawed off every poplar limb within three feet of the ground, leaving a distinct browse line. The only hare we saw in the valley near the den was immobilized in the jaws of a red fox that loped by us one evening. Arctic ground squirrels, common in all but dense spruce stands throughout Kluane, were not to be seen here. In the stands of aspen parkland a few miles to the north, ground squirrels could be measured in tons of animals per square mile. We discovered this from a week's study, using live capture-recapture trapping. But the wolves were primarily eating moose, and probably this produced a more favourable ratio of energy spent in capturing food to energy gained in eating it than if their diet had consisted mainly of small mammals.

We spent three days with the wolves in their pristine retreat and briefly watched the four pups rolling and playing together. But the adults (there were at least three of them) were secretive and spent most of the time in the spruces near the den. It was a thrilling few days, listening to the howls of different individuals — bass, tenor, long and drawn out, short and yappy. But it ended sadly.

One evening we howled and there was no reply; the wolves were gone. While our presence may have hastened their move, we had been careful to play by the rules and had left them completely alone most of the time. They may have moved despite us; the pups were five or six weeks old, normally time for them to leave the natal den. The day after they left, however, one pup was back. It was alone. The others may have forded a fast mountain stream which it couldn't cross, or perhaps it lacked stamina and lagged behind even before they reached the fast water.

We were witnessing the cruel side of natural selection, the very process that has made the evolution of advanced forms of life, including man, possible on this earth — the pruning and culling necessary to keep every population of every species viable and healthy. In a way, it was arrogant to interfere. The abandonment was part of the answer we sought to the causes of pup mortality — not just random death but the selective process. Wolves, like almost all species of animals, produce far more young than necessary to maintain constant numbers. This is not evidence that evolution is frivolous or wasteful, but rather that a high birthrate is necessary to allow selective mortality to take place. Man, the humanitarian, is one of the few who deviate from this pattern of high birthrate followed by high selective loss, unbalancing the equation by propping up the poorly adapted and inferior, and consequently ensuring their genes a continuing place in his gene pools. How could it be any other way? We could hardly toss all babies out in the snow to see which could survive. But by sidestepping a basic life process, will our humanitarian tendencies ultimately lead to biological failure?

The following morning the pup was singing a death song — longer howls than when it was part of the chorus — but there was no reply. It would be dead in a week, but for much longer than that we have continued to hear its plaintive cries. We could do nothing. We might have trapped a few snowshoe hares for it, but even that would have been difficult there. Even when winter came, it would still have had no chance. We could have caught the pup and turned it into a captive wolf, but I have worked long enough with captives to understand that wolves belong to the wild; to separate the two destroys the wild quality which is the essence of both. The mountains were indifferent: they just threw back the sound. The goats glued to the rocky crags cared no more than if they were papier-mâché. The rest of the pack, wherever they were, were out of hearing. But we listened with a lump in our throats. Because for us, that pup's howls were not merely calling out to other wolves to save it; they were crying out to man to save the species, to stop harassing it with traps and snares and guns; pleading for us to save some wild places in our mad quest for resources to satisfy our irresponsible population growth; begging for land where wolves and their kin can continue to play out their wild drama in peace.

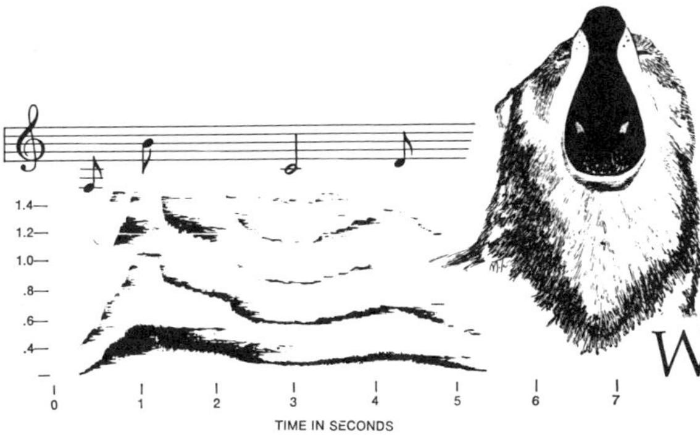

9 Wolf Music

Sweet song of the wilderness, wild, free, and haunting, striking a chord on the memory-strings of our evolutionary past, plucking a note of fear from shadowy recollections of our childhood picture books, trumpeting adventure with overtones of buffalo hunts and mountain men and tales dimmed by time — old Two Toes, Bigfoot, the Custer Wolf — all long dead. From a mile away on a still night, whispering a sad reminder of carcasses strewn around a poison bait; from a hundred feet sounding a grim alarm that if man's relationship with other living things is too coarse to allow compassion for the wolf, man is singing his own death song.

Few sounds in this world evoke such a range of emotions. For some, wolf music is as deeply moving as Beethoven's Fifth Symphony reverberating in the chambers of a great concert hall. For others, it is a call to arms, stirring up mad hate, blind terror. It both symbolizes the modern movement to preserve some wilderness and mocks man that he hasn't conquered it all. Each person's response is dictated by unconsciously formed impressions gained by chance references to wolves in books, newspapers, and movies.

But wolves know nothing of all this. Alone on a night's hunt, some deep-seated wolf emotion builds up until, unable to resist any longer, the animal throws back its head and howls. Vocal chords buried under the coarse hair of its throat vibrate as the air from its chest is pressured outward by tracheal muscles. Sound waves oscillate from its open mouth, and the wild music is thrown out on the night air. No thought of man. A wolf howl is simply part of the biological essence of a wolf, as real as its curved canine teeth and steel-grey eyes.

Wolves have stalked this earth for a few million years, and while howls cannot be dug up with other fossil remains, we know enough of the anatomy of the wolves' ancient forebears to surmise that their music rang through the valleys in the west even before the final uplift of the mountains was complete. The functions of canine teeth and eyesight are obvious. But what about howls? They, too, have been tested by hundreds of thousands of years.

To discover why wolves howl, we must first believe there is an answer. According to the Darwinian theory of natural selection, essentially all shapes and sizes, colours and sounds, have a biological function. All characteristics of living matter, plant or animal, have been moulded and fashioned by selection, and have provided in the past, or provide now, an advantage to the species. A child's question 'Why?' concerning any living thing is as profound as if asked by a professor emeritus, and upon the ability to ask this simple question and derive an answer revolves the whole framework of modern biological science.

Only a few years ago we thought that birds sang just to make us happy. Not until 1920 did Eliot Howard teach us that by singing, birds were proclaiming territories, a shocking blow to our egocentricity. The story has continued to unfold, and we now know something about the subtleties of bird song — that dialects appear in different geographical localities just as they do in most languages, that neighbouring birds of the same species sing differently, and that one bird can recognize another by its song. Even the domestic fowl, which we have selectively bred for centuries to improve its meat but not its intelligence, gives different types of alarm calls: danger from the air evokes calls different from those caused by danger on the ground.

But we know even less when we turn to mammals. Elk bugle, moose bellow, whales squeak and buzz, squirrels chatter, mountain lions scream, and wolves, most musical of all, sing.

For two years I asked every wolf I heard why it sang. I picked that 'why' because it was deeply fixed in my mind from night after night of census routine. I picked it also because only by building up an understanding of all the time-tested adaptations of a summit predator might man, another summit predator, perceive the worth of wolves and alter his treatment of them accordingly.

During the censuses I had accumulated anecdotal information that, while disjointed and in need of a common theme, formed a first impression of the musical score. The uncanny ability to pinpoint and come great distances to the source of sound, demonstrated by the three wolves at Fork Lake that misty morning, perhaps is fundamental to any social structure in wolf packs. Without it, wolves could communicate only through vision or scent. Over distances of a mile or more, pack cohesion may be impossible without a form of vocal communication. Gregarious predators in the Order Carnivora are usually vocal — wolves, coyotes, jackals, cape hunting dogs; more solitary species are normally silent — Arctic and red foxes, and most of the cats.

Another observation, made at the rendezvous sites, was that spontaneous howling by lone wolves was associated with separation of individuals from the pack. The nights when most of the adults left the rendezvous site were often filled with song; other nights when nobody went hunting were invariably silent.

Other factors besides separation affected willingness to howl. Time of year was one: wolves responded more frequently as the summer progressed. Weather was another; if any rain, even the slightest drizzle, was falling, wolves rarely howled. The length of time elapsed from a previous howl was yet another factor. Ethologists recognize a time lag between responses in many forms of animal behaviour, and liken this to the time required to refill a toilet tank with water. In animals, 'action-specific energy' must build up before they will respond again.

The more I glimpsed the musical score, the more questions seemed to arise. Can wolves identify each other by their howls? Is information coded in howls, and if so, what is it? Does some particular part of a howl trigger a response?

In answer to this last question I knew that wolves answer even poor human imitations of howls. No two humans howl in exactly the same way, but all seem equally successful in attracting answers from wild wolves. Nor must sound always be particularly wolflike. One summer I inadvertently conditioned a pack of wolves to respond to the slam of a truck door. My nightly procedure at the time was to stop on the highway at the same spot and howl. I parked on a slope, often slamming the truck door when I got out. Eventually my howl was superfluous — the slam of the door was enough to start the pack howling.

While these isolated experiences were intriguing, they were not sufficient to provide any detailed knowledge of wolf howling. To learn when and why wolves howled meant getting to know them much better.

The opportunity to do so was provided by three captive wolves at the Wildlife Research Station in Algonquin Park. While I owe these animals an apology for being party to inflicting the indignity of captivity on them, two facts can be raised in partial defence. First, they had never known complete freedom and, accustomed to both humans and wolves, they appeared contented with their lot. Secondly, studied and written about by others, most notably Doug Pimlott, they have graciously contributed to a new understanding for their species. That does not in any way excuse me, and my work with these three wolves has sensitized me to feel sorry any time I see a tame or captive wolf.

The first summer I studied only one of the three, named Big Grey, and transported him three miles north of the Research Station to an isolated area where he would rarely hear other wolves. I built a pen for him in a small clearing behind an abandoned ranger's cabin, and in mid-May began my research. By playing howls of wolves with whom Big Grey had grown up, as well as howls of wolves he had never heard, I hoped to find some differences in his behaviour, howling or otherwise, that would show he recognized individuals by their howls. By comparing Big Grey's behaviour when he howled with characteristics of his howl, I hoped to find differences in the sound that might convey information. By his response to various frequencies, I hoped to discover if a particular note or combination of notes would trigger a response.

But Big Grey refused to howl. All through May and June, night after night at half-hour intervals, I played tapes or howled myself from the cabin window, out of sight of the wolf, but heard no replies. Then, at the end of June, Mary joined me at the cabin. On her first night there she howled. Big Grey responded immediately. I could not believe it. She howled again. He answered. They repeated the performance all night, but if I played a recorded wolf howl or howled myself, Big Grey was silent. Big Grey did not know Mary and had never heard her howl before. Yet Mary's howl worked.

The next night we tested Big Grey by howling identical notes, C above middle C on the piano. Again Big Grey consistently responded to Mary, but not to me. So, no triggering frequency was involved in stimulating his reply. Although we never found out why he was so enthralled by Mary's howl, we did learn how well he could discriminate.

The next day I recorded Mary's howl, and that night played the tape out of the cabin window. Big Grey did not answer. Thereafter, he consistently showed that he could distinguish Mary's live howl from a recording of her howl. What was the difference?

I stood by the pen and she tested me. (Instead of howling back, I merely noted my conclusions.) Big Grey proved more consistent than I in distinguishing the sounds.

The next winter we replaced a wolf as the receiver with a piece of electronic apparatus, a sound spectrograph, which makes a graph of any sound played into it. It portrays not only the main sound, known as the fundamental frequency, but also the harmonics. Harmonics produce the quality in a sound. By differing in number and intensity, they make a note played on a guitar sound different from the same note played on a piano. The sound spectrograph showed us that one of the ways Mary's taped howl differed from her live howl was by a minor difference in these harmonics. The tape recording of her howl stressed the second harmonic rather than the first, whereas, in her live howl, the first harmonic was stronger than the second.

For a wolf to possess the ability to distinguish subtle differences in harmonics argues that this ability has some adaptive significance, that it is actually used. It could be used if wolf howls had similar differences in harmonics, and if these differences conveyed some sort of meaning.

My problem, then, became analogous to learning a foreign language without a dictionary or instruction. First, you must recognize units of sound. Then you must determine the pattern in which they occur. Finally, you must recognize the situation in which they are given.

If the English language were analysed in such a way by someone unfamiliar with it, he would discover three facts about it: there are certain units of sound; certain sounds, or words, occur in patterns; and certain words occur in particular contexts — as when your hammer drops through a pane of glass you have just replaced.

One night, one of the captive wolf pups at the Research Station seriously injured its foot in the wire of its pen and had to be destroyed. All through that night, over and over again, an adult wolf housed in an adjacent pen gave a very melancholy but beautiful howl. It rose to a high note, D above middle C on the piano, held this for about two seconds, then dropped abruptly four or five notes, followed by a second abrupt drop. In all, it dropped more than a full octave. I had occasionally heard the same howl given by wild wolves, always by single animals far from any other replying voices. Later, when another of the adult wolves was isolated from his pen-mates, he often uttered this sorrowful-sounding howl.

Single, sudden drops in pitch occurred in many howls we heard from both captive and wild wolves. I chose these, along with characteristics of the beginning and the end of howls and the pitch of the highest notes, as the units of sound for my analysis.

That first summer, I tape-recorded more than 250 of Big Grey's howls, adding another 400 howls from all three wolves the following year. Where possible, I noted the behaviour of the howling wolf — whether it was lying down, standing up, pacing, or running in its pen.

Gradually, as information accumulated, a pattern emerged. After hours of examining sonograms of recorded howls I discovered that in most cases it was possible to distinguish individual wolves by their howls. Despite a great variation in pitch, length, and other features of each wolf's howls, patterns in the harmonics were distinct. Big Grey's howls, for example, always contained at least five harmonics; the other two wolves normally had only two, and never more than four harmonics. These two wolves, Dagwood and Scamp, differed mainly in their first harmonic. Dagwood's first harmonic was usually stronger than Scamp's. In addition, Scamp showed a unique feature in all his howls: when the pitch of the fundamental frequency was between the notes B and F-sharp, the first harmonic was strong. Most of the midsections of his howls fell within this range. At the beginning and end of his howls, however, this harmonic faded out.

In short, the properties of the howls of individual wolves differ, as do the singing voices of humans. Now we knew that wolves can detect subtle differences in sound, and that they have distinctive howls. The probability that they do actually identify each other by howls appeared greatly strengthened. Just as some birds, such as the white-throated sparrow, recognize and react differently to songs of neighbours and strangers, wolves, too, may differentiate pack members from intruders by howls. Howling almost certainly plays a role in passive territorial defence, acting to delineate boundaries with as little strife as possible. Without individual recognition, pack members might well have many more direct encounters with animals in other packs, thus defeating a prime benefit of a territorial system.

When I examined the features of howls in relation to the wolves' behaviour, I found that two of the wolves emitted different howls when they were pacing from when they were standing still. When he was pacing, Dagwood's howls almost never contained a sudden drop in pitch, which they always did when he was still. The exact opposite was true of Scamp. Big Grey's howls did not vary with his movement.

Other differences in the howls of individuals appeared related to whether the howls were stimulated by us or occurred spontaneously. When spontaneous, Big Grey's howls differed in almost every measurable way from howls that were stimulated. Whereas stimulated howls rose gradually in pitch, spontaneous howls began with a sudden upward jump in pitch, often as much as five notes, and ended with an abrupt downward slur; they were normally higher in pitch throughout and shorter in length.

The transfer of information can take place on two levels — universal and individual. Communication that is universal (occurring within the species in general) requires a symbolism that is the same throughout the species. Communication on the individual level may occur among animals that learn to recognize individual traits in animals with which they associate.

The wolves I studied had grown up with one another. Like two humans of long acquaintance who communicate by means of familiar gestures, I discovered that in certain situations my three wolves could transfer information — such as the identity of the wolf, and perhaps its emotional state — by varying the sound.

Big Grey hated to be left alone. In August of the first summer we suddenly realized that he had learned to associate our walking past his pen in a particular direction with isolation. We normally left him for a period of hours every four or five days while we returned to civilization for supplies and to backpack in more dog food for him. The trail out went by his pen. We made our discovery one day when, instead of continuing along the trail, we stopped to pick blueberries. Big Grey began to howl. The trail ran along a creek most of the way, and we realized that we rarely had had a chance to hear him howl after we left the cabin. He howled every thirty seconds or so for the next hour. We carefully circled back to the pen and watched him pacing rapidly back and forth between howls.

The next day we repeated the situation and he howled again. From then on we were able to record many howls when he thought we had left him. It was these spontaneous howls that, on later analysis, proved to be different from howls stimulated by Mary.

This discovery, while it may seem obvious, adds an important link to understanding at least one reason why wolves howl, and what kind of information howls may be conveying. If something akin to what humans call loneliness is felt by wolves, and if loneliness causes howling, wolves separated from pack members will howl. It does not matter what we call the emotion. To be coldly scientific we can call it emotion X. But if one manifestation of emotion X is howling, the by-product is communication: information transmitted about the location of the wolf, perhaps its identity, and, if the howl was initiated by emotion, perhaps reflecting this too, as did Big Grey's howls.

There is much more to wolf howling than this. In his classic study of wolves in Alaska, Adolph Murie cites many examples of wolves howling when he approached. Group howling, too, is another category. These howls may be motivated in an entirely different manner.

While there is still much listening to do to crack the full information code of wolves, man is listening as never before — thousands are listening. It started in 1963 on an August evening, when eight hundred campers, mostly urban people on vacation, met for a 'wolf-listening' night in response to posters at the museum in Algonquin Park. They travelled in cars, stopping every mile or so while the naturalist howled. For many this was the first time they had listened to the night forest: a loon on a far-off lake, the deep chorus of bullfrogs. Then, at one stop, they heard a wolf. Suddenly they were no longer standing on a paved road leaning against a modern car. They were no longer mere spectators. For a brief few minutes, they were accepted back into the wilderness.

Park naturalists now conduct regular howling nights in Algonquin Park, and interest is growing fast; about five thousand people take part in these trips every year.

These people will never stand back and see the wolf legislated out of existence. They will hear that howl and in spirit leave the city and return to the wilderness any time the word wolf is mentioned. For whatever the wolf howl means to another wolf, it means more to us.

10
Where the Wolf is White

The white wolf stood motionless, outlined against the rock-strewn landscape, scarcely fifty feet away. Ears erect, nostrils flared to identify the strange scent, the wolf was glimpsing (probably for the first time) the form of a man.

The wind on the great Arctic wilderness on Baffin Island had erased all traces of my footsteps, and now, at the crest of a low hill, I faced the animal I had travelled over a thousand miles to find.

Carefully, keeping my eyes on the wolf, I reached in my pack for the camera. Too late. The animal wheeled abruptly and, with loping stride, vanished over the rise.

It was the summer of 1965, and, under contract to the Canadian Wildlife Service, I was to determine the feasibility of studying wolves in this treeless part of North America, a study which would allow me and other biologists to observe predators interacting with their prey.

Except for a scattering of scrubby spruce on its southern tip, the vegetation on Baffin Island does not grow more than a few inches high. Miles of bell heather, dark green among the rocks, is mixed with mosses and dwarf willows which are hardly more than spreading vines. In wetter areas the ground is covered with sedges and low grasses.

We had gathered at the Frobisher Bay airport a week before my first encounter with the white wolf. A strong summer wind was whipping the Arctic poppies growing along the runway, on that brilliant July day. Our party consisted of Andrew Macpherson, then Chief Mammalogist for the Eastern Region of the Canadian Wildlife Service and a veteran in Arctic experience; Gaylen Armstrong, a biologist employed by the Wildlife Service, and myself. Later we were joined by two Eskimos, John Kilabuck and Tom Tukpani. Our plan was to fly over the central and southern portions of Baffin in a chartered flying boat and prepare a sectional census of caribou populations. When we located a concentration of these animals, we would land at the nearest possible point and set up camp. Later in the summer, when the study was completed, the plane would return and pick us up. For my part, I hoped to find wolves by staying close to the caribou, their chief prey in this northerly region.

For some days the plane flew back and forth in wide sweeps across the flat expanse of tundra with the regularity of a ploughman turning new furrows in a limitless field. Two freshwater lakes, each exceeding a thousand square miles, provide variety in an otherwise regular terrain. Nettiling Lake, astride the Arctic Circle, is dotted with so many islands that its eastern shoreline is difficult to define. The Great Plain of Koukdjuak sprawls over six thousand square miles, its flatness broken only by low snaking eskers, once the gravelly stream beds of glaciers. In places, thousands of snow geese speckled the tundra, or wheeled in loose flocks, like snowflakes whisked off spruce branches by a gust of wind.

On the east coast the scene changes. Deep fiords divide mountains crowned with ice caps — all that remains of the vast glaciers that once covered the island. In the central interior, rocky ridges sweep in undulating waves between yellow-green sedge meadows. Here we saw caribou — hundreds of them, grazing in the wet meadows or trotting with their stiff-legged gait over the ridges.

We set up camp at Fox Bravo, an abandoned outpost on the Distant Early Warning Defence Line. An airstrip where planes no longer land, some small buildings falling into ruins, a litter of empty oil drums, a tangle of steel beams — only these spent toys of civilization remaining to rust for decades in the great Arctic wilderness. Four hundred miles from Frobisher Bay — miles of rock, lakes, turbulent rivers, and impassable muskeg. Two hundred miles north of the Arctic Circle. Fox Bravo was a forgotten link in the famous DEW Line, now too remote even for Eskimo hunters.

We piled our gear on the edge of the airstrip and waved goodbye to the pilot. There we were, five men, far away from all other human beings in one of the remaining truly remote wilderness areas of this continent.

The next day I began my search for wolves; a search that involved walking miles, scanning the tundra from each rise, and examining each sandy stream bank for tracks. When I spotted evidence of wolves, I set out scent posts in an effort to dupe the animals into showing themselves.

A scent post has the opposite effect upon wolves from what it has on humans — it attracts them. A good *set*, as it is called, consists of a short, specially treated stick placed in a cavity between some rocks or in a small hole dug out of a bank. The stick is treated by dipping it into a bottle of 'perfume' which is irresistible in its powers of animal attraction. Every trapper has his own jealously guarded formula. The brand I used, although its ingredients are no secret, required a strong stomach to compound; it is made up of skunk oil, wolf urine, beaver castor, and rotted venison, which, at the time, had been fermenting for over two years.

The first week I established half a dozen scent posts within a five-mile radius of our camp, and each day I made my rounds. In my searchings I began to sense the subtle interplay between the living and non-living in this harsh land. On the summits of the hills were whole lessons in Arctic ecology. Here, luxuriant grass grew taller and thicker than on the lower slopes. Rocks on the crests of these hills were frequently coloured with an orange lichen seldom found elsewhere. These were bird perches where snowy owls scanned the country for lemmings. Their droppings enriched the earth with nitrogen, an element in critically short supply in most tundra soils. The vegetation responded, and this attracted seed-eaters like horned larks and pipits, which added more nitrogen to the soil, further speeding up the process of enrichment. Arctic foxes paused at these hilltop oases, checking for nesting snow buntings. And wolves too had visited them, marking boundaries, sniffing the vegetation, reading the olfactory messages of those who had been there before them, and leaving behind a record of their own passing.

I crossed patterned ground where the freezing and thawing action of surface soil had moved huge rocks into circular designs. I walked through fields of mud-boils where sub-surface soils had seeped upwards through cracks to the surface and smothered the vegetation. I stood on hillsides where the soils were creeping downwards, like frosting on the side of a cake, moving imperceptibly, a few centimetres each year. This process, called *solifluction,* enriches the lowlands while robbing the upper slopes of essential nutrients. With no life-giving soil, the rock-viscera of the land are exposed — in some places huge rough-edged boulders strewn about just where they had dropped out of the retreating ice; in other places, fields of fragmented granite chewed up and half-digested by the glaciers.

In forest lands to the south, the living zone stretches from twenty feet or so down in the ground, where the roots of pines reach out for soil nutrients, to a hundred feet up in the air, where olive-sided flycatchers perch on spruces, searching the treetops for insects. But on the tundra of Baffin Island, all life is squeezed into inches, offering a never-ending fight for existence between the winds that blow unobstructed a few hand-spans above the earth's surface and the permanently frozen ground less than a foot below. Only when geese fly over, or longspurs spiral high in the air during their courtship flight, does the band of life exceed these narrow limits.

July days are long in the Arctic, and sometimes my search continued past midnight. My companions were shooting a representative sample from the caribou herd to assess its condition. Measurements were taken, fat deposits carefully weighed, and information collected on age and reproductive success. All this was done in an attempt to understand why the caribou population on Baffin Island appeared to be remaining constant, while that on the nearest Arctic mainland was dwindling. The meat was not wasted; it was laboriously carried back to camp, to be distributed on our return among the needy people at Frobisher Bay.

Gaylen saw the wolf first, spotting it while scanning the hills from our boat. It was pure white, standing out clearly against the greens and browns of the tundra. Before we were able to land, it disappeared from the top of a rocky ridge a mile away. By the time we reached the hill, the wolf reappeared and came loping straight towards us. It veered off, then stopped. My heart was pounding as I reached for my camera. Would anyone believe that we had found an Arctic wolf so easily if I didn't get that picture? But the wolf was concerned about other things we had yet to discover. It turned and galloped out of sight.

It is one thing to see a wolf; it's another thing to keep contact with it long enough to provide information of scientific value. As I considered our next move, the animal reappeared about a quarter of a mile away, joined by two more wolves.

The three stood facing us, stark white against the green heather and howling in chorus. So reluctant were they to leave that we were certain their den must be nearby. We found four pups at the base of a rock wall, only a few yards away. Several minutes passed before they realized that their well-concealed home had been discovered. Then, overwhelmed by curiosity, they climbed the rocks to get a better view of us. The wind had dropped, and we stood motionless. The only things that moved were mosquitoes — in two black clouds, shrouding our heads. The boldest pup came closer and closer. Then, suddenly, inquisitiveness gave way to fear and it bolted in panic, triggering alarm in the others. Perhaps their fear resulted from some slight movement, or the click of my camera shutter. Possibly the pup was reacting to a new experience and would have run from its first acquaintance with a lemming or an Arctic hare. In a flash the adult wolves raced after the pups which, by the time they were overtaken, were far out on the plain.

The den was actually a tunnel into the wall, with two entrances, one higher than the other. In front of it was a well-trampled sedge meadow and a small pond. The wolves had lived there for at least two weeks, judging by the bones and droppings scattered around.

The white wolf of Baffin Island is one of twenty-four subspecies of North American wolves which taxonomists originally described. Many of these have long been exterminated. This northern race, *Canis lupus manningi*, was little known until 1938 when T. H. Manning collected several white wolves while exploring Baffin's west coast. The wolves of the high Arctic Islands and Greenland are white, or nearly so, although there is some individual variation — shades of grey or black which may not be distinctive in the field. The Baffin Island wolf is perhaps the most uniformly white of all; those we saw showed only a tiny black spot on their tails, a characteristic common to all wolves and coyotes. The reason that Arctic wolves are white isn't fully known. The white wolf is easily seen against the summer tundra, and though it blends with its surroundings in winter, this camouflage does not seem to be of any protective benefit, since the white wolf is the prey of no other animal. But colour can be an advantage, and it may be a help to the Arctic wolf in its hunting during most months of the year, when the ground is white.

The next day Gaylen brought me back to the den, which was eight miles from our base. I had decided to camp there for a few days, to see if the wolves would accept my presence and go about their normal activities. This trip turned out to be a misadventure.

When we first appeared near the den, fortune was still in my favour. My field notebook for Tuesday, August 3, reads:

'While we were eating lunch on a ridge within sight of the den, I heard a faint howling. Down on the plain below were the three adult wolves and two of the pups. They were trotting southeast, towards the den, with a graceful gait, tails flowing behind. The lead wolf suddenly saw us; we were silhouetted against the sky. They all immediately swung in our direction, the lead animals making a semicircle around us at about seventy-five yards while the others moved up the ridge farther back, trotting a few steps, stopping to look at us, trotting again. They all circled back to the plain, deciding we were not worth further concern, and, strung out in a line, continued towards the den. When almost there, the lead wolf stopped, and in a cluster together, they howled — pups too — short treble howls, weakened by the wind.

'At the den, two adults stood together in the wet grass, while the other walked along the wall of rock, exploring the crevices. The pups followed. Then, in a curious display, they all came together, nose to nose, and milled around each other for about fifteen seconds. As they broke their huddle, they started back along the plain, retracing their route, strung out again, but this time moving faster. The pups lagged behind. Occasionally the adults stopped to let the pups catch up. As they passed us, they paused to look up, but only momentarily. Soon they were far out on the plain and out of sight.'

What had they communicated to each other in that nose-to-nose display? They did not howl. Were less audible sounds made? Did scent, or some subtle movement, perhaps a posture, convey a meaning? Similar displays by wolves had been observed before, on Isle Royale in Lake Superior, but the significance of this behaviour remains as yet another mystery about which we can only wonder.

The first sign of trouble showed up before Gaylen left to return to the base camp. I had forgotten my tent poles, a matter of concern where there were no trees to be found within hundreds of miles. I improvised by using my pack to prop up the front of the tent, and a large boulder to raise the back. This arrangement, which resembled a spent parachute, left about a foot of clearance inside; room enough, if I lay flat on my back. My only retreat from the bloodthirsty mosquitoes was here, and I hoped that a good strong wind would come up and clear them away. As it turned out, I hoped too well.

Gaylen returned to the boat and disappeared across the lake. I was alone with the ringing calls of golden plovers and Baird's sandpipers, with the nodding Arctic cottongrass, and the hordes of mosquitoes — and the wolves. A short-tailed weasel popped up by my tent, looked me over, then dived back in among the rocks. Seconds later, as though it couldn't believe its eyes, it popped up again, and ran off. Then the wind began to rise and the calls of the plovers sounded distant. I crawled into my low-slung tent and lay there looking out. Spread around me was the simplest of the world's terrestrial ecosystems, similar to the types found in deserts of shifting sands, which are characterized by short food chains, impoverished soils, and very little diversity of plants and animals. Simple as these ecosystems may be, they remain environments of mystery. Their very simplicity attracts ecologists who hope to discover in them fundamentals of how all ecosystems function. But the questions are many and the answers few. Why do lemmings cycle in abundance, reaching their highest population densities approximately every four years? Why do ptarmigan populations have a ten-year cycle? Scientists agree that population fluctuations are not universally caused by weather, disease, predators, or even available food supplies; the effects of all these have been studied, but then discarded because they rarely offer complete explanations of the changes in numbers. Possibly more common is the existence of a self-regulatory mechanism, sometimes brought about by an automatic control of birthrates through socially imposed restraints. Or perhaps population densities are geared to some essential nutrient which may periodically be in short supply? Could populations rise and fall as a result of changes in the creature's inherent ability to withstand the rigours of a harsh environment? These are some of the current suggested explanations, none of which have yet been fully proved true.

Wolves present other questions. Why haven't these predators increased in number to a point where they jeopardize the very survival of caribou? Why do wolves not starve during the denning season, if the migratory caribou leave before the pups are old enough to travel? Considering the great advances in knowledge made in this scientific age, it is surprising how little man knows about even simple ecosystems.

The second indication I had of trouble ahead appeared in the early evening, when clouds formed and blocked out the lowering sun. The wolves had not returned and I was becoming concerned that they had moved away permanently. The clouds increased, spread across the sky by a strengthening wind. I crawled out of the tent and fastened down the sides with more rocks. Before I could squirm back under the canvas, the storm hit. Sheets of water, driven almost horizontally by the violent gale, threatened to carry away my whole camp. This was no passing shower or pleasant mosquito-chasing breeze; it was two days before the sun reappeared.

Textbooks say, 'The Arctic is a frozen desert, with rainfall comparable to other arid regions of the world.' That would have seemed humorous to me as I lay in my prone position, peering out at the deluge, but, at the time, my sense of humour was, like my clothes, thoroughly dampened. My pack, which supported the roof of the tent, rubbed against the canvas sides. Water poured in and accumulated on the floor. It wasn't long until I was in the first stages of submersion, and numb with cold, since the temperature had dropped to about five degrees Celsius.

Just about the time dawn would have been breaking a thousand miles to the south, I began to shiver uncontrollably. In self-defence I lit my small primus stove and placed it as far as possible, about twelve inches, from the side of the tent. As it sizzled and spluttered in the wetness, I amused myself by puncturing holes in the floor at strategic spots to drain out some of the water. This brought on the next disaster. As I was channelling water to one of the holes, it jumped my binocular case dyke and ran towards the primus. I grabbed the stove. The flame hit the side of the tent, and — whoosh — I was staring at grey sky. The tent had caught fire, a gaping hole had been burned in its side, and the flames had been blown out, all in a matter of seconds. I climbed through the hole and assessed my plight. There was more water in the tent than out. The wind was so strong that I had to face away from it to catch my breath. I walked towards the wolf den but it seemed to be deserted. There was no break in the storm-filled sky and my gear was soaked — camera, notebook, spotting scope, and food. I had been close to success, and now the whole enterprise had been wiped out so thoroughly that there was nothing else to do but to hike back to the base camp.

The tundra was still writhing in agony under the cold, wind-driven deluge. Along the rocky ridges, mountain avens vibrated rhythmically on their inch-high stalks. The poppies lashed to and fro on wiry stems. A ptarmigan flew up at my feet; the wind caught its wings as it sailed over a hill with primary feathers tightly folded and tail askew. The caribou, plentiful on the plains the day before, were nowhere to be seen, having found shelter among the rocks where the wind was less severe. In places the sedge meadows were knee-deep in water; the streams, which a few days ago had been trickling run-offs, were now torrents.

When I reached the base camp, I found my companions busy scraping caribou bones and trying to carry on a conversation by outshouting the noise of the canvas slapping against the tent frame. They seemed not the slightest bit surprised that I had returned, a fact that eased my pride but did little to lessen my disappointment at missing an opportunity to watch the wolf pack.

Two days later the rain stopped altogether. Once again caribou trotted over the ridges and the plovers called. The cottongrass lifted its head. But the wolves came back to their den only occasionally. They appeared, now and then, at the base camp and ate a few free meals of caribou meat before we discovered them and raised our stores well out of their reach. The pups had moved, and we never saw them again. Occasionally, wolf prints showed up around my scent posts, along with white hair where the animals had rolled on the scent.

The wolves appeared to pay little heed to the caribou, and, strangely, the caribou seemed little concerned with the wolves. Only once, when a large bull and a wolf met unexpectedly as they both came over a hill, was there a short chase, but the caribou easily outdistanced his enemy.

By the middle of August, the sunny nighttime hours had been replaced by twilight. I had accomplished my purpose: I had found wolves at their den, and assessed the possibilities of further studies of their behaviour among their prey. As the caribou sampling was also satisfactorily completed, we were ready to return to civilization.

Before we set out from Frobisher Bay, Andy had arranged for the aircraft to pick us up at Fox Bravo, and, when that day arrived, most of our gear was packed beside the airstrip. The weather was bright and sunny, and we didn't expect problems. Had we been more perceptive, we would have been concerned about two pieces of news which we had picked up the night before on our portable radio receiver. Through a scramble of splutters and crackles we had learned over Frobisher Bay radio that the Otter aircraft detailed to bring us out had been forced to make an emergency landing. The second piece of news which might have given us cause for concern was that the senior government administrator, with whom Andy had arranged for our pickup, had been posted to Ottawa. Word of the emergency landing had not disturbed us, since other aircraft were stationed at Frobisher Bay. The transfer of the administrator seemed unimportant, as we assumed that others had instructions about our pickup.

The day passed — no aircraft. And the next day, and the next. It was my job to stay close to the runway and meet the plane when it arrived, while the others resumed their caribou sampling. These days were not wasted. My ears, although attuned to the drone of an airplane, heard instead the restless wind carrying fragments of longspur song, the harsh call of the parasitic jaegers as they tossed in the sky currents, and the plaintively beautiful cry of the Arctic loon. The loon, which was nesting on an island in a small lake near the airstrip, sailed about the pond with kingly grace, the snowy whiteness of its head flashing mirror-like against the blue water. In a few weeks it would begin its unusual migration, following the ancient sky-road of the species straight west along the Arctic coast all the way to the Pacific Ocean, before turning south. For some two thousand miles, this curious route cuts across almost at right angles the path of all other migratory birds.

One morning, while scanning the eastern sky for the plane, I saw an Arctic fox trotting agilely over a rock-field. Not much bigger than a large house cat, it moved quickly, some purpose locked in its canine mind. Later, when Gaylen returned to camp, he reported seeing the same fox, shortly after I had spotted it. It was feeding on the remains of a caribou the Eskimos had collected a few days earlier. While Gaylen was observing the fox, a wolf appeared and galloped towards it. The fox looked up, and realizing that the master had arrived, took off at full speed. The wolf did not pursue the fox; its very appearance had been enough to assert its superior rank in the animal hierarchy. But then the wolf spotted Gaylen. Reluctantly it turned and trotted up the ridge, pausing every few yards to look back at the only creature which, its instinct warned it, should be left alone.

A week had gone by, and still no plane. The weather was beginning to turn cold and we realized that something must be done. Over a bowl of caribou soup we discussed our plight and agreed upon a plan. It was decided that Andy and John would leave next day to walk overland about sixty miles east to the nearest manned defence station, called Fox 3. Their way led over the rough tundra broken by countless small lakes. By carrying only a rifle and no provisions, they planned to take as little time as possible. Gaylen, Tom Tukpani, and I were to remain at Fox Bravo and, by means of flares, signal any planes that might be flying over this section of the DEW line.

We ferried Andy and John across the lake and watched them disappear over the rolling tundra. The day was cool and cloudy, and snow was not far off. We had been back at camp only a few hours when a plane, a DC3, droned into sight high over us. As I pointed the flare skyward, it suddenly struck me that our predicament was becoming melodramatic. I released the spring. Pop — and a small orange ball shot up a paltry ten feet into the air, extinguishing itself almost immediately. It reminded me of the flash fire in my tent. So much for flares! The pilot would have missed seeing my signal even if he had been watching with binoculars from the end of the airstrip.

We put in the next two days scraping hides and caribou bones; working helped pass the time. On the evening of the second day, during a period of audibility, we received the news from the Frobisher radio station, four hundred miles away, that Andy and John had made it to Fox 3. Andy had sent a message from there back to Frobisher Bay, where it had been put on the air for our benefit.

Two days later, after a delay by bad weather, a DC3 flew in low over the hills and landed at Fox Bravo. We took down our tents, stowed our gear aboard, and were airborne for home. In a moment we were up over the lake with the Arctic loon riding its glassy surface, over the rock-field where the fox had run from the wolf, over a band of caribou grazing near the wolf den. We left Fox Bravo and its wolves to the solitude of the Arctic wind.

For thousands of years, that wind sang its lonely song in the north, in boundless theatres crowded at various times by glaciers, vast seas, or dry land. The song sweetened as tundra soils were formed, fed from below by abraded rock and from above by decaying vegetation — first, lichens and mosses; later, grasses and a host of flowers. Then the wildlife moved in — the lemmings and Arctic hares, and a variety of birds, and the wolves, all adding their songs to the music of the wind until it swelled into a great chorus.

But with the white man came the first discordant notes. His music — the shuddering, grinding, raucous noise of bulldozers, the screech of jet engines, the whine of drilling rigs, the shattering clamour of snowmobiles, the earth tremors of gigantic explosions — doesn't harmonize with the ageless chorus.

Is the irreplaceable music of the Arctic crashing towards a finale of discordant chaos? Oil, gas, copper, lead — all these rich resources are needed by man. But what of the wildlife, the beauty, the remoteness? Is there no way to make man's activities harmonize with nature's?

11
Wolves, Wilderness, and Man

It is just a typical weekday morning. The air lanes are streaked with jets, catapulting people across the continent. The urban freeways are humming — six lanes each way. On monuments above the streets, commuter trains charge the cities from every direction. Steel shrieks on steel as subway trains cut time into fractions. People rushing headlong. Time is money. Build another airport. Multiply the traffic lanes. Speed. Without it the country can't progress!

But out beyond metropolitan anywhere, unseen by the crowds, who are enveloped in a haze of unawareness, a flock of pectoral sandpipers rises from the mudflat of a northern lake. It is not moving nearly as fast as the jet; nevertheless it is moving. Energy is not modern man's invention; it continues to move up an age-old pathway, from the cell protoplasm of simple organisms eaten by the birds to metabolic pathways that provide the power for their migration.

And somewhere else a wolf lopes across the gravel bar of a braided mountain stream, slow in comparison to a 350-cubic-inch automobile engine, but still moving. Within the animal, nitrogen that once formed part of the protein in a deer is being transformed into ammonia and urea, ready to be excreted and returned to the soil. There it will be rebuilt by soil bacteria before it begins another trip through living cells; a trip that is slow, but essential to all life.

And somewhere else again, a bumblebee hums away from a wild columbine; slow compared to the commuter train, but still able to get the job done — the job of transferring pollen from anther to stigma to fertilize a new generation of green plants, upon which all living things, including man, depend.

But we have never fully realized the immense costs that modern technological society levies on the environment.

It is Saturday afternoon. Mobs cram the shopping mall and the world looks great. Shop windows are filled with merchandise; stores blaze with light. Air conditioners hum from their perches near the mall's glass roof. Heels clack on coloured cement tiles. The artificial trees spaced here and there create an elegant illusion for the few thousand people of wandering through a clear-cut plastic forest. Why, I remember when this *was* just a forest!

But around the corner, hidden by a few hundred miles of power lines and a ground fog of indifference, giant machines ten storeys high are ripping open green rolling hills, gutting the land square mile by square mile to provide coal to light those shop windows, coal to make the air conditioners hum. Or is it oil that runs this shopping centre? Oil from offshore drilling rigs with spills every few weeks, or tankers that collide in fog-shrouded harbours. Oil that is nonrenewable. Where did the sand come from for the coloured cement tiles, or the plastic for the trees? Everything on display is a product of the land. Somewhere a cost has been levied; fewer minerals, of which the earth has but a finite supply; a little less clean air; a little less photosynthesis; a little less oxygen.

It is any weekday evening. In the glare of television lights, a political leader walks up to a battery of microphones. The chatter in the press box ceases. The politician clears his throat, tilts his head to what his wife says is his best angle, and smiles into the camera. 'Ladies and gentlemen,' he begins. 'Your government has decided to put up for immediate tender the construction of what will be one of North America's, indeed the world's, greatest engineering feats of water diversion. With boldness and foresight we will harness the wasted water falling on two hundred thousand square miles of wilderness and put it to work for the betterment of mankind.'

And as he speaks, beyond his vision even on the elevated platform, a flock of Canada geese lift off the 'wasted water' on their way to fascinate hundreds of school children from Northern Ontario to Texas with their precise and graceful flight. A trout splashes the surface, a challenge to the fisherman for whom a few days spent in this wasteland is an annual highlight. The wind sighs through spruce needles, opening their pores just a little to speed transpiration. The balance of water in the air, in the ground, and in living cells alters slightly in a million micro-habitats in the spruce forest. But there are natural limits to this water balance that have maintained the spruce forest for thousands of years before man ever evolved.

We have nailed up our sign all over the earth: 'Under New Management — by Man, not Nature.' Nature is tolerant: for a while it adapts. But over how much of the earth can we trade biological cycles for bulldozers, ecological diversity for engines, plant productivity for pavement? When do the ledgers turn red? Time will be the final cost accountant.

Nobody knows when life began. Indeed, differences of opinion about what constitutes life would prevent a final decision even if the record were clear. The beginning of life happened about two billion years ago, give or take a few hundred million. Before that, for almost as long again, our planet spun lifeless in its present position in space. But time spans of that magnitude are far beyond human comprehension. If only we could fully grasp the meaning of time, and could visualize the ages, we might feel more reverence for today's green world and the vital ecological processes on which it depends. If we were aware of our great good fortune, we would modify considerably our human activities — not by returning to a civilization of loincloths and wooden ploughshares, but by realizing that in satisfying our multiple needs we must tread more lightly upon this earth. We only have one earth: if life exists outside our galaxy it is too remote to be any help.

As far as we know, the right combination of atmospheric gas from which life could evolve occurred on our planet only once. At a fleeting moment in time, the right combination of atmospheric gases (ammonia, hydrogen sulphide, hydrogen cyanide, methane) and energy sources (electrical discharges, exploding meteorites, or ultrasonic vibrations, nobody is sure which) resulted in simple organic compounds, the precursors of life. It is ironic that the first uncertain steps towards life were taken in an atmosphere devoid of oxygen — an atmosphere that would be lethal to us today.

Among the compounds synthesized may have been amino acids, the structural parts of protein. We can synthesize them in a chemistry laboratory from elements which scientists believe may have been present in the earth's primitive atmosphere. But we can only be amazed and thankful that the synthesis ever happened on its own.

How self-reproducing matter managed to establish itself is even less clear. One step may have been the formation of giant molecules able to incorporate into themselves substances from their environment, and later split into replicate halves. In theory that is possible, but whether it actually happened this way will probably always remain a speculation. When scientists examine the continuity of life — the chromosome split, the polarization of halves, the chemical bonding of homologous chromosomes from each parent — they find the detailed processes so orderly that they are predictable. The molecular structure of genes has been thoroughly unravelled. But we only have a glimmer of comprehension as to how it all began, or why it all works. We can only stand back in amazement, and with perhaps a growing humility.

Now the curtain is drawn back and we glimpse a fragment of our immediate past. Light falls on only the last ninth or so of the earth's history — the most recent 500 million years. At the farthest known point we read the saga of our existence in fossils laid down in Cambrian seas. The evolution of living things is the product of a combination of phenomena: selection, speciation, genetic drift, mutations, laws of probability, and population genetics among them. The marvel is not life as we know it, but the processes of evolution that made it.

Thousands of chance occurrences played their part in evolution. At one stage an ozone layer, an essential insulator around the outer edge of the earth's atmosphere, evolved. That this layer could form may have been inevitable, given the atomic weight and structure of oxygen, from which it is made. But that fact only transfers our good luck from the layer itself to its atomic structure. Ozone reflects the sun's ultraviolet radiation, and not until it took its place could life creep from the seas without frying.

First, reptiles explored this newly available terrestrial environment. Later, between 255 and 185 million years ago, birds and mammals evolved as two of many branches from a basic reptilian stock. From this stock, some reptiles developed fish-like tails and returned to the sea. Others evolved forelimbs adapted by a leathery membrane for flying. Still others became bipedal, and one group, leading to mammals, developed specialized teeth and structurally stronger lower jaws. It was as if nature was experimenting to discover which forms of life were best suited to the changing world. Only those whose adaptations were useful survived. While we know this sudden diversification occurred, we don't know why. It may have taken place during a period when the sun bombarded the earth with cosmic radiation, resulting in a rapid rate of genetic alterations. But this hypothesis advances very little our explanation of how and why so many species came into being, and only transfers our wonder to the physics of the sun and why the solar activity occurred.

Mammals first appeared about 160 million years ago, or in the often-used comparison of evolution to a twenty-four-hour clock, in the last hour and a half. Primitive wolves and other canidae showed up much later — between fifty and twenty-five million years ago. Their fossil record begins in North America, and that they did not reach Africa, the 'cradle of man', until the evolution of primitive man was well advanced may have been another stroke of good fortune. It allowed primate evolution, one branch of which led to man, to advance in the absence of a potential enemy. Zinjanthropus, the earliest true man and father of us all, left evidence that he lived 1.7 million years ago. Doubtless he was forced to compete with the canidae, which held several of the best cards in the evolutionary deck: greater speed, better scent ability, night vision. But early man had become bipedal, perhaps through pure luck and a series of genetic mutations — nobody knows exactly why, and with forelimbs freed for hurling rocks and spears, he was able to hold his own. It was lucky that he had become able to use hands in addition to feet before the wolves arrived!

Have we been thankful in these 1.7 million years? How Zinjanthropus regarded nature is not known: we can assume, however, that he had little means of destroying it. But just considering the Christian Era, the last two thousand years, have we been thankful? Reverence for nature is part of what we consider primitive religions. Modern religion is man-oriented. The Ten Commandments say nothing of an ethical regard for anything but our fellow man. We have told ourselves we are made in God's image, and have gone forth to 'multiply and subdue the earth'. It is wrong to kill a human being. It is right, even commendable, to kill a whole living ecosystem. The man with blueprints in his hand for grandiose schemes of land development is seen as a visionary. Rarely in the past has the burden of proof for the necessity of his possible felony been on his shoulders. And he violates neither criminal nor religious laws. But now, very lately, there has been a new awakening to the danger. Whether it has come in time to stop the momentum is still in doubt.

Only seventy-five elements made our entire world, living and non-living, us included. We see them lined up on the chart in any high school chemistry classroom, as if putting them in neat rows is the best man can do with them. That they combined to create the beauty and order of this earth is the most profound, humbling, and spiritually uplifting truth in the universe. Yet we build cathedrals to find God, and after church go out to check the pole trap behind the barn.

A reverence for our thin mantle of life, spawned out of unexplained events which took place in the past, is man's most compelling need today.

What is the solution? We must eat; therefore much of the land must be cultivated. We need natural resources such as wood, sand, and gravel with which to build homes; considerable land must be set aside for that. We need places to put those homes, and means of transportation, and energy to provide fuel for both — requiring more land. We wish to possess recreational and cultural opportunities, so we build beyond the basic physical needs — still more land is needed. But all this land need not be written off. The philosophy in the phrase 'design with nature' is packed with significance. But we must have wilderness or at least relatively unaltered land to run the world's physical plant — to produce oxygen, regulate the hydrological cycle, recombine and move nutrients, maintain soil productivity, provide ecological stability, and to fulfil some of man's spiritual requirements. When enough people become convinced that it is just as important to retain land for wilderness as it is to put it to other uses, the welfare of living things on our planet, man included, will begin to improve. Survival will be ensured when the conviction is translated into action: when we halt world population growth, reduce our material demands, and on a world-wide scale co-ordinate our efforts towards long-range planning for the best use of land — action motivated not only out of fear for our survival but because we have finally acquired a philosophy of reverence for nature.

How much wilderness do we need? Nobody can answer that. Yet for any country this must be a most vital issue, and one on which we cannot afford to make a mistake. With only our present fragmentary knowledge of the natural processes that support life, some very broad guidelines can be stated. We need enough green landscape to prevent an oxygen deficiency; enough natural water flow to maintain a relatively constant proportion of atmospheric water, ground water, and surface water; large enough parcels of wild lands to ensure the survival of representatives of all species alive today — our genetic savings for an uncertain future. But since some people will never understand these things and can only see land as so much real estate to be bought, sold, and developed, we can be sure that to save enough wilderness will require all the aggressive action, both personal and political, that we can command. The key to any success will be a concerted effort to understand our physical environment more fully. If we work towards that goal as we have for reaching the moon, then someday life scientists will take their place at the head of the table at every land-planning conference. Today we can do little more than sound a warning.

Wolves, and the questions of their impact on their prey, the way they regulate their numbers, and how they communicate, are significant not so much by themselves as by being part of the matrix of life. But wolves may offer man some useful perspectives on himself. After all, we are both predators, both highly developed socially, both with family units as the basis of our societies, both communicating with facial expressions and by vocal means. Both may even have similar emotions. Wolves can tell us some things now, if we care to listen: that face-to-face ritualized aggression may often avoid outright physical aggression, which should indicate to us that much conflict in human life may be avoided if we communicate with each other directly; that killing members of your own species is a human trait, normally absent even in animals man considers cruel; that joy and exuberance for life are not exclusively human emotions. In time, wolves may tell us how to be more successful in regulating our numbers. By watching them interacting with natural environments we may find ways to overcome the egocentricity that has seemingly placed us above nature. For wolves and wilderness are inseparable. In another but very real sense, man and wilderness are inseparable.

The wind has dropped; the moon has yet to lift over the pines. The Big Dipper and the North Star stand out clearly, pointing the way. Across the black water floats the distant howl of a wolf. The pulse of man quickens, as it has ever since we began to compete directly with wolves over a million years ago. But today the howl reminds us that our past is deeprooted in wildness. It warns us not to go too far in destroying natural environments. It epitomizes the wilderness we have fought so successfully to conquer and now must fight to save.

Epilogue

Wolves and Science

The wolf has been the subject of scientific research for a variety of reasons. Among these are its conflicts with man, its contributions to the general scientific theory of predation, its endangered status through most of its present-day range — particularly the United States — and more recently our wish to see it survive in our wilderness areas. Wolves once lived throughout North America from the high Arctic to the Mexican desert, and in Europe and Asia north of the Tropic of Cancer. Now they have gone from southern Canada and from most of their former range in the United States, where they are found only in northern Minnesota and Alaska. In the many areas from which wolves have disappeared, their extinction is the result of an alteration of habitat and ruthless persecution. Wolves are more vulnerable to man than the closely related coyote with its greater reproductive capacity and its elusive near-solitary character. Wolves still live in Russia, Poland, Spain, Portugal, Italy, northern China, and India, though they have gone from Britain and are rare in Germany, Hungary, and the Scandinavian countries.

Modern research into wolves has shown, with sad irony, that many of the preconceptions that have caused man to persecute them are mistaken. If a wolf kills a deer, there is not necessarily one less deer available for a hunter, except in the short run. Paul Errington, a distinguished American biologist, who devoted most of his life to the study of predation in the wild, was one of the first to question such simple predator-prey relationships. He questioned, too, the general applicability of conclusions from studies of laboratory populations of unicellular animals and invertebrates. These studies showed that predators have the capacity to limit prey numbers, and that, in turn, prey numbers limit predator populations. But the observations had somehow been expanded into generalizations that were not borne out by most of the recent field studies. In his book *Of Predation and Life* (1967), Errington wrote:

> Apart from a number of extreme or dramatic cases of predation depleting prey populations in ways that are self-evident, my inclinations are to look very critically upon figures presented by themselves as proof of population effect. They may constitute no proof at all, however imposing they may be when superficially regarded.

Behind this statement was Errington's argument that many species of prey can compensate for losses through greater production. Also, many animals that die as a result of one cause, such as predation, would soon have died anyway from other causes. That is why many species can withstand the annual harvest by man through hunting or trapping without depletion of their breeding stock.

While, as a generalization, Errington's belief that predators rarely limit prey numbers is probably much closer to the truth than the earlier hypothesis, a variety of relationships may exist between predators and prey. Errington recognized this too when he said that 'diversity and complexity in these interrelationships are wholly consistent with diversity and complexity in the forms of living things'. Wolves and their prey demonstrate that a diversity of relationships does in fact exist. In 1944 Adolph Murie, who studied wolves in Alaska, concluded that wolves were limiting numbers of Dall sheep and caribou. In turn, the relative scarcity of these species was limiting wolves. This finding was similar to the results previously mentioned that came from laboratory research, except that no cycle of numbers was implied by Murie's results. Similarly, the results of an extensive wolf control program that wiped out wolves in northern British Columbia in the mid 1950s resulted in a dramatic increase in moose, and thus indicated that wolves had been limiting moose numbers. But on Isle Royale in Lake Superior, a recent reinterpretation of fifteen years of data by David Mech, one of the principal biologists involved, indicated that wolves had not been limiting moose. This came to light when the moose population increased from 600 to 1,000 despite a dense and protected wolf population.

Other biologists have also found that wolves were not limiting prey: Ian McTaggart Cowan, studying elk in Banff National Park, Alberta, in 1947; Daniel Thompson, studying white-tailed deer in Wisconsin in 1952; Frank Banfield studying caribou in the Canadian Arctic in 1954.

Thus, it is apparent that no single relationship exists between wolves and their prey populations. More than that, a single generalization is even impossible over extended periods of time in the same place. Our research in Algonquin Park has shown that relationships between wolves and white-tailed deer may change — they are dynamic. When deer were plentiful, before about 1963, Douglas Pimlott concluded that, while wolves were a major cause of mortality, they did not seem capable of reducing deer numbers, on the basis of the amount of meat required to sustain the population. The wolf population was stable despite food resources that seemed capable of supporting more wolves. Apparently some self-regulatory mechanism in the wolf population was preventing its further increase. Then deer numbers dropped, part of a decline that has occurred concurrently across the whole northeast extreme of the species' range, whether or not wolves were present. In the two places where wolf studies were conducted, Algonquin and Minnesota, the wolf population has also declined. As well, it has adjusted its food habits; whereas once 80 per cent of their diet in Algonquin was deer, my two assessments in 1971 and 1974 showed this had dropped to 30 to 40 per cent, with beaver and moose replacing deer. However, indications are that despite these adjustments wolves in Algonquin may now be cropping as many deer or slightly more than are replaced by the annual production of fawns. The relationship of predator and prey has changed.

Evolution is on-going; ecosystems are constantly changing; adaptation and adjustments continually take place. From evolution's standpoint, one relationship between predator and prey is no better or worse than another; they are all part of the process of natural selection that keeps animals in tune with their environment. Tested by time, all predator-prey relationships contribute to the perpetuation of life — hunter and hunted.

Scientific research into wolves has shown that wolf-prey interactions vary with species of prey, with the place, and with the time; that wolves often cull out inferior prey animals; that wolves are rarely wanton killers; that wolves have the capacity to limit their own numbers; and a host of descriptive aspects of their lives such as the existence of rigid social hierarchies and extended periods of play. There are still important questions left. One is whether wolf predation is compensated for by a reduction in other causes of mortality, so that even where data superficially indicate that wolves are limiting prey they may not be in fact. Nowhere in Canada, Alaska, or Minnesota, where control is being exercised on wolves, has this critical question been answered. It is hard to design an experiment to settle this, since the question is really whether a particular animal would have lived to reproduce successfully if it hadn't been killed; since the future success of a dead animal cannot be determined, solid evidence is lacking. The question might be answered in specific areas where predators are controlled — but only if there were an adjacent area where no control was exercised, so that the fate of the prey populations in the two areas could be compared. But unfortunately predator control programs are almost always politically rather than biologically motivated.

A second intriguing question is how and why wolves sometimes regulate their own numbers. Social exclusion of old or young, reduced breeding because of the hierarchies in wolf packs, are possible explanations. Perhaps the mechanisms are diverse too, like predator-prey relationships; future research may tell.

All these complexities about the biology and role of wolves in ecosystems leave most biologists who study wolves with a profound respect for the species. While it is unfortunate that pat answers are not available to allow more accurate generalizations about wolves, I share a hope with other students of wolf ecology that, in spite of the gaps in our knowledge, man is wise enough now to recognize the rights of wolves to share the wild places we have left.

. . . the illustrations include